Medical Intelligence Unit

Immune Mechanisms in Allergic Contact Dermatitis

Andrea Cavani, M.D., Ph.D.
and
Giampiero Girolomoni, M.D.
Laboratory of Immunology
Istituto Dermopatico dell'Immacolata, IRCCS
Rome, Italy

Landes Bioscience
Georgetown, Texas
U.S.A.

Eurekah.com
Austin, Texas
U.S.A.

Immune Mechanisms in Allergic Contact Dermatitis

Medical Intelligence Unit

Eurekah.com
Landes Bioscience

Copyright ©2005 Eurekah.com
All rights reserved.
No part of this book may be reproduced or transmitted in any form or by any means, electronic or mechanical, including photocopy, recording, or any information storage and retrieval system, without permission in writing from the publisher.
Printed in the U.S.A.

Please address all inquiries to the Publishers:
Eurekah.com / Landes Bioscience, 810 South Church Street, Georgetown, Texas, U.S.A. 78626
Phone: 512/ 863 7762; FAX: 512/ 863 0081
http://www.eurekah.com
http://www.landesbioscience.com

ISBN: 1-58706-209-7

While the authors, editors and publisher believe that drug selection and dosage and the specifications and usage of equipment and devices, as set forth in this book, are in accord with current recommendations and practice at the time of publication, they make no warranty, expressed or implied, with respect to material described in this book. In view of the ongoing research, equipment development, changes in governmental regulations and the rapid accumulation of information relating to the biomedical sciences, the reader is urged to carefully review and evaluate the information provided herein.

Library of Congress Cataloging-in-Publication Data

CONTENTS

EDITORS

Andrea Cavani, M.D., Ph.D.
and
Giampiero Girolomoni, M.D.
Laboratory of Immunology
Istituto Dermopatico dell'Immacolata, IRCCS
Rome, Italy
Chapters 5, 6, 10

CONTRIBUTORS

Cristina Albanesi
Laboratory of Immunology
Istituto Dermopatico dell'Immacolata, IRCCS
Rome, Italy
Chapter 10

David A. Basketter
Safety and Environmental Assurance Center
Unilever Colworth
Sharnbrook, Bedford, U.K.
Chapter 1

Stefan Beissert
Department of Dermatology
University Münster
Münster, Germany
Chapter 11

Frédéric Bérard
Unité Immunologie Clinique et Allergologie
Lyon, France
Chapter 4

Tilo Biedermann
Novartis Research Institute
Department for Allergic and Inflammatory Diseases
Vienna, Austria
Chapter 9

Ponciano D. Cruz
Department of Dermatology
The University of Texas Southwestern Medical Center
Dallas Veterans Affairs Medical Center
Dallas, Texas, U.S.A.
Chapter 13

Marie Cumberbatch
Syngenta Central Toxicology Laboratory
Alderley Park
Macclesfield, Cheshire, U.K.
Chapter 3

Rebecca J. Dearman
Syngenta Central Toxicology Laboratory
Alderley Park
Macclesfield, Cheshire, U.K.
Chapter 3

Andrea Dötze
Max-Planck-Institut für Immunobiologie
Freiburg, Germany
Chapter 2

Bertrand Dubois
INSERM
Unité Immunologie Clinique et Allergologie
Lyon, France
Chapter 4

Alexander H. Enk
Department of Dermatology
University of Mainz
Mainz, Germany
Chapter 7

Katharina Gamerdinger
Max-Planck-Institut für Immunobiologie
Freiburg, Germany
Chapter 2

Nicola Gilmour
Safety and Environmental Assurance Center
Unilever Colworth
Sharnbrook, Bedford, U.K.
Chapter 1

Christopher E.M. Griffiths
Dermatopharmacology Unit
The Dermatology Centre
University of Manchester
Hope Hospital
Salford, Manchester, U.K.
Chapter 3

Richard W. Groves
Academic Medicine and Therapeutics
Imperial College of Science, Technology and Medicine
Chelsea and Westminster Hospital
London, U.K.
Chapter 3

Sven Hellwig
Max-Planck-Institut für Immunobiologie
Freiburg, Germany
Chapter 2

Whitney A. High
Department of Dermatology
The University of Texas Southwestern Medical Center
Dallas Veterans Affairs Medical Center
Dallas, Texas, U.S.A.
Chapter 13

Brandon G. Howell
Department of Dermatology
Johns Hopkins University School of Medicine
Baltimore, Maryland, U.S.A.
Chapter 8

Dominique Kaiserlian
INSERM
Unité Immunologie Clinique et Allergologie
Lyon, France
Chapter 4

Ian Kimber
Syngenta Central Toxicology Laboratory
Alderley Park
Macclesfield, Cheshire, U.K.
Chapters 1, 3

Tadashi Kumamoto
Department of Dermatology
The University of Texas Southwestern Medical Center
Dallas, Texas, U.S.A.
Chapter 12

Adam J. Mamelak
Department of Dermatology
Johns Hopkins University School of Medicine
Baltimore, Maryland, U.S.A.
Chapter 8

Hiroyuki Matsue
Department of Dermatology
The University of Texas Southwestern Medical Center
Dallas, Texas, U.S.A.
Chapter 12

Norikatsu Mizumoto
Department of Dermatology
The University of Texas Southwestern Medical Center
Dallas, Texas, U.S.A.
Chapter 12

Akimichi Morita
Department of Dermatology
The University of Texas Southwestern
Medical Center
Dallas, Texas, U.S.A.
Chapter 12

Mark E. Mummert
Department of Dermatology
The University of Texas Southwestern
Medical Center
Dallas, Texas, U.S.A.
Chapter 12

Francesca Nasorri
Laboratory of Immunology
Istituto Dermopatico dell'Immacolata,
IRCCS
Rome, Italy
Chapter 5

Jean-François Nicolas
Unité Immunologie Clinique
et Allergologie
Lyon, France
Chapter 4

Grace Y. Patlewicz
Safety and Environmental Assurance
Center
Unilever Colworth
Sharnbrook, Bedford, U.K.
Chapter 1

Martin Röcken
Department of Dermatology
and Allergology
Ludwig Maximilians University
of Munich
Munich, Germany
Chapter 9

Pierre Saint-Mezard
Unité Immunologie Clinique
et Allergologie
Lyon, France
Chapter 4

Daniel N. Sauder
Department of Dermatology
Johns Hopkins University School
of Medicine
Baltimore, Maryland, U.S.A.
Chapter 8

Claudia Scarponi
Laboratory of Immunology
Istituto Dermopatico dell'Immacolata,
IRCCS
Rome, Italy
Chapter 10

Agatha Schwarz
Department of Dermatology
University Münster
Münster, Germany
Chapter 11

Thomas Schwarz
Department of Dermatology
University Münster
Münster, Germany
Chapter 11

Silvia Sebastiani
Laboratory of Immunology
Istituto Dermopatico dell'Immacolata,
IRCCS
Rome, Italy
Chapter 6

Camilla K. Smith Pease
Safety and Environmental Assurance
Center
Unilever Colworth
Sharnbrook, Bedford, U.K.
Chapter 1

Akira Takashima
Department of Dermatology
The University of Texas Southwestern
Medical Center
Dallas, Texas, U.S.A.
Chapter 12

Hermann-Josef Thierse
Max-Planck-Institut für Immunobiologie
Freiburg, Germany
Chapter 2

Binghe Wang
Department of Dermatology
Johns Hopkins University School of Medicine
Baltimore, Maryland, U.S.A.
Chapter 8

Hans Ulrich Weltzien
Max-Planck-Institut für Immunobiologie
Freiburg, Germany
Chapter 2

PREFACE

Allergic contact dermatitis (ACD) is a very common T cell-mediated disease of the skin with a high socioeconomic impact in both professional and nonprofessional settings. The large interest on cutaneous reactions to haptens has involved in the last twenty years not only dermatologists and allergists, but also immunologists and basic researchers. The main reason for the attention focused on ACD is that it has represented, and still does, one of the most reliable and informative models for understanding dendritic cell and T cell biology. Compared to other T cell-meditated skin diseases, such as psoriasis and atopic dermatitis, ACD has valuable animal models which have provided a large body of information on its pathogenesis. Furthermore, ACD can be easily induced and, importantly, it can be monitored to track the kinetics of the immune response. After being considered for decades a classic $CD4^+$ T cell-mediated delayed type hypersensitivity reaction, many differences with T cell responses to proteins have emerged. ACD is currently considered a prototypic $CD8^+$ T cell-mediated disease. Indeed, a reasonable overview of the pathogenesis of ACD should take into account the notion that its development, expression, and regulation strictly depend upon a complex cross-talk between resident skin cells and immigrating leukocytes, including neutrophils, dendritic cells, monocytes and T lymphocytes. Among the latter, $CD8^+$ T cells certainly have a critical effector role in the expression of the disease. However, amplification of the disease is strongly promoted by the release of pro-inflammatory cytokines and chemokines by type 1 $CD4^+$ T lymphocytes. Of great importance is the characterization of T cell subsets which modulate immune reaction to chemicals and which may represent a formidable tool for new therapeutic approaches to ACD and other T cell-mediated skin disorders.

Andrea Cavani
Giampiero Girolomoni

ABBREVIATIONS

ACD	allergic contact dermatitis
APC	antigen presenting cell
CHS	contact hypersensitivity
CLA	cutaneous lymphocyte-associated antigen
CTL	cytotoxic T lymphocytes
CTLA-4	cytotoxic T lymphocyte associated antigen-4
CGRP	calcitonin gene-related peptide
CTACK	cutaneous T cell-attracting chemokine
DC	dendritic cells
DNCB	dinitrochlorobenzene
DNFB	dinitrofluorobenzene
DNP	dinitrophenyl
DTH	delayed-type hypersensitivity
ELC	EBV-induced molecule 1 ligand chemokine (CCL19)
GM-CSF	granulocyte macrophage colony stimulating factor
GRO	growth-regulated oncogene
FasL	Fas ligand
FITC	fluorescein isothiocyanate
HEV	high endothelial venules
KO	gene-targeted knockout
ICAM-1	intercellular adhesion molecule 1
ICE	IL-1β converting enzime
IFN	interferon
IL	interleukin
IL-1RA	IL-1-receptor antagonist
iNOS	NO synthase
IP-10	IFN-induced protein 10 (CXCL10)
I-TAC	IFN-inducible T cell alpha chemoattractant
LC	Langerhans cell
LF	lactoferrin
LFA-1	lymphocyte function-associated antigen-1
LLNA	local lymph node assay
LN	lymph node
LT	lymphotoxin
mAb	monoclonal antibody
MC	mast cells
MCP-1	monocyte chemoattractant protein-1
MDC	macrophage-derived chemokine
Mig	monokine induced by IFN-gamma

MIP	macrophage inflammatory protein
MHC	major histocompatibility complex
MMPs	matrix metalloproteinases
NO	nitric oxide
OX	oxazolone
PMN	polymorphonuclear granulocytes
PSGL-1	p-selectin glycoprotein ligand 1
RANTES	regulated upon activation, normal T cell expressed and secreted
SLC	secondary lymphoid tissue chemokine
SOCS	suppressor of cytokine signaling
STAT	signal transducer and activator of transcription
TARC	thymus- and activation-regulated chemokine
Tc	T cytotoxic
TCR	T cell receptor
TGF	transforming growth factor
Th	T helper
Tr	T regulatory
TNCB	trinitrochlorobenzene
TNBS	trinitrobenzene sulfonic acid
TNP	trinitrophenyl
TNF	tumor necrosis factor
UCA	urocanic acid
UV	ultraviolet
VCAM-1	vascular cell adhesion molecule1

Chapter 1

Approaches to the Predictive Identification and Assessment of Chemical Contact Allergens

David A. Basketter, Grace Y. Patlewicz, Camilla K. Smith Pease, Nicola Gilmour and Ian Kimber

Summary

The prospective identification of potential contact allergens and their subsequent safety assessment is the pivotal activity in successful management of this risk to human health. Although much can be learned from the chemical and physical properties of a substance, the definitive information in respect of sensitising hazard/risk derives from an assessment of the integrated response of the immune system. In recent years, the focus for such assessments has begun to switch from the guinea pig to the mouse, notably to the local lymph node assay (LLNA). The utility of the LLNA for hazard identification is now accepted and its regulatory status defined. Once a potential contact allergen has been identified however, the vital clue to accurate safety evaluation is the assessment of the potency of the allergen. This can be achieved using the LLNA and the data employed in safety evaluation of chemicals. In addition, the data can be deployed in a practical manner into new processes in regulatory toxicology. Ultimately, the aim must be to avoid completely the use of animals for identification and assessment of potential skin sensitisers. Although some basic strategies exist for hazard identification, the more profound challenge will be to understand how in vitro endpoints correlate with sensitising potency to the point where the data can be used in risk assessment.

Introduction

Allergic contact dermatitis (ACD) results from the T-lymphocyte mediated immune response to a chemical allergen that comes into contact with the skin (Fig. 1). The small allergenic molecule, often referred to as a hapten, penetrates the skin and binds to a carrier protein typically by a covalent bond to form an allergenic hapten-protein complex. Skin enzymes frequently play an essential role in activation and/or inactivation of chemicals during this earliest stage of the response. The hapten-protein complex is processed by antigen presenting cells, principally the dendritic Langerhans cells (LC) of the epidermis. These cells migrate to the draining lymph node, where they present the chemical to T lymphocyte to provide the stimulus for antigen-specific commitment and the production of memory and effector T lymphocytes. Subsequent contact with a sufficient dose of the chemical will then result in the expression of the clinical signs of ACD.[1] The primary elements associated with the induction of ACD therefore involve skin penetration, the possible metabolic activation of unreactive chemicals and covalent binding to skin protein. Understanding of these elements has developed rapidly in

Immune Mechanisms in Allergic Contact Dermatitis, edited by Andrea Cavani and Giampiero Girolomoni. ©2005 Eurekah.com.

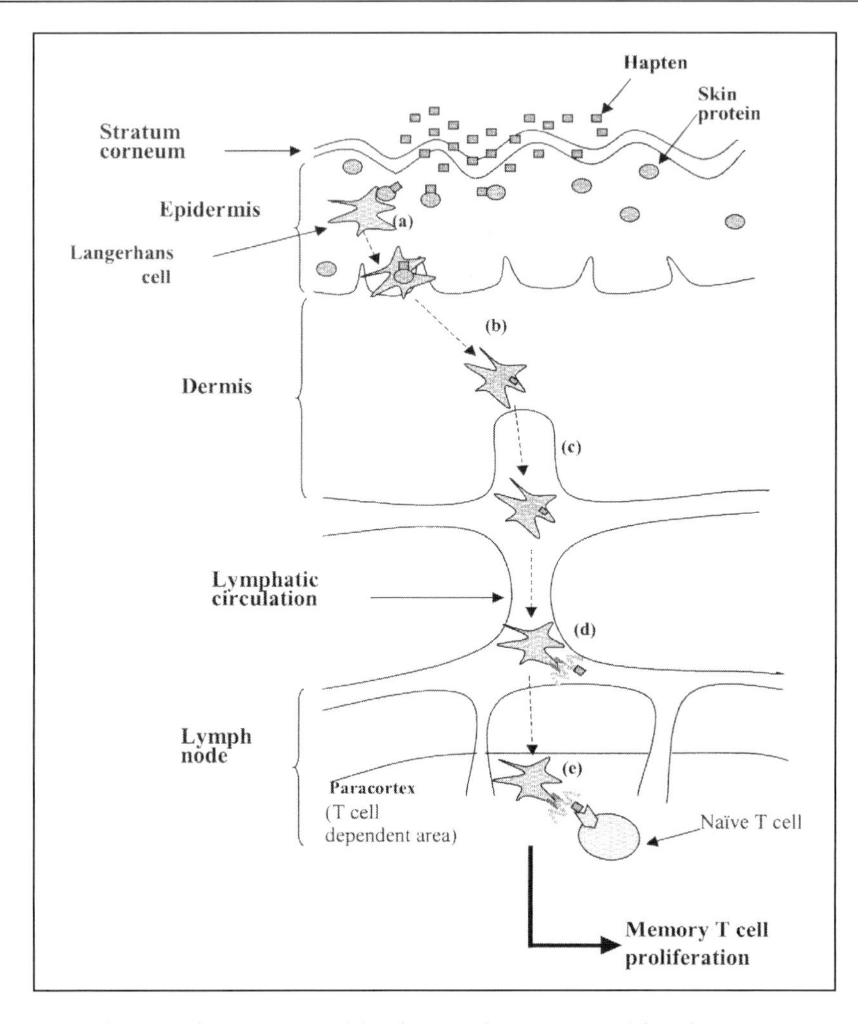

Figure 1. Mechanisms of sensitisation in delayed-type IV hypersensitivity. This schematic represents the current theory of the processes involved during the sensitisation phase of ACD. A hapten (square) is first absorbed into the epidermis, where it can bind to skin protein thus forming an immunogen. These modified-proteins may then be recognised and internalised by a LC (a). The LC then migrates from the epidermis into the dermis (b) and finally to the draining lymph node (c), whilst maturing into a dendritic cell. The LC processes the immunogen into peptides (c) that 'carry' the hapten, which is bound covalently, and display them on their surface (d). The peptide-hapten complexes are then recognised by the TCR of a naïve CD4$^+$ T cell residing in the paracortex of the lymph node (e). This recognition, then stimulates the generation and proliferation of a population of memory T cells.

recent years, providing opportunities for greater prospective identification of potential contact allergens, as well as aiding their recognition in relation to clinical diagnosis. In this chapter, we provide an overview of the state of knowledge in this area.

Skin Penetration

The first hurdle a chemical must cross in order to behave as a contact allergen is presented by the stratum corneum. Some of the physicochemical properties that modulate penetration into the viable layers of the epidermis are fairly well characterised: hydrophobicity (as measured by the log octanol/water partition coefficient, logP), the presence of charged groups, aqueous solubility and molecular shape and size.[2] All of these properties can be measured and/or calculated for a substance of defined chemical structure. For example, chemicals with a low molecular weight (< 500 g/mol) and with a log P of >1 indicating they are relatively hydrophobic are very likely to be able to penetrate the lipid-rich stratum corneum effectively. In contrast, larger hydrophilic, charged molecules will penetrate very poorly (reviewed in ref. 2). However, the value in being able to make predictions related to skin penetration of chemicals only becomes clear when allied to an understanding of what the chemical might do once in the viable epidermal layers and thus in contact with LC. In addition, a potentially compromised skin barrier (effected by physical or chemical insults or an inherent abnormality e.g., ceramide deficiency) will have a substantial impact on the overall rate of penetration of a chemical into the epidermis.

Chemical Characteristics

There is a long established connection between the ability of chemicals to react with proteins to form covalently linked conjugates and their skin sensitisation potential.[3] For skin sensitisation to occur, once a chemical has penetrated, it must be able to partition into and between the relevant intra-, extra- and sub-cellular compartments of the epidermis, in order to be appropriately and sufficiently bioavailable. Then the chemical, or its metabolite, must be sufficiently electrophilic to react covalently with nucleophilic groups on skin protein to produce the complete antigens capable of stimulating the immune system. This reaction with protein is likely to be selective for particular amino acid units depending upon the chemical functionality of the sensitising chemical.

Uncovering the mechanisms of skin sensitisation provides valuable insights into the behaviour of chemicals for hazard identification purposes and subsequent risk assessment.[1,4] Currently, predictive testing involving animals has to be performed with the ultimate aim of identifying sensitisation hazard and potency and reducing the risk of producing ACD in man. An alternative approach involves examining key physicochemical properties of substances and relating these to sensitisation potential. Such structure activity relationships (SARs) provide a means of investigating and predicting the toxicological effects of chemicals. SARs are based on the principle that the toxicological properties of a chemical are dependent upon the chemistry of the toxin of interest.

One approach to establishing SARs is the Relative Alkylation Index (RAI), a mathematical model derived by Roberts and Williams,[5] based on electrophilicity and hydrophobicity parameters as well as the dose of chemical. The RAI model has been used to evaluate data on various sets of skin sensitising chemicals (reviewed in ref. 6).

Other approaches include the development of empirical quantitative SARs (QSARs) by application of statistical methods to sets of biological data and structural descriptors. For many years, guinea pig tests have provided the data from which SARs have been derived. Enslein et al[7] developed QSAR models for assessing dermal sensitisation using guinea pig data for 315 chemicals. Two models were proposed; one for aromatics (excluding chemicals with 1 benzene ring) and the other for aliphatics and chemicals with 1 benzene ring. The models resolved a qualitative potency in terms of bands of classification. Rather than a hypothesis based approach, a variety of descriptors were computed for the chemicals selected; stepwise 2 group discriminant analysis was used to build the models and identify relevant descriptors. An optimum prediction space algorithm was incorporated into the model to ensure predictions were only made for new chemicals within the model domain. This model was incorporated into the TOPKAT (Toxicity Prediction by Komputer Assisted Technology) expert system. Though a useful predictive tool, the model is constrained by the limitations of the (Guinea Pig

Maximisation Test) GPMT assay on which it is based (and which itself over-predicts the response), the scope of the training dataset and because previously identified modes of action were not considered.

The introduction of the LLNA (Local Lymph Node Assay—see below for more details) with its quantitative endpoint for skin sensitizing potency has proved a major boost to skin sensitisation QSAR. The LLNA involves topical application of the test chemical to mouse ear skin followed by quantitative measurement of the T-cell proliferation response in the draining lymph node, assessed as a function of the incorporation of tritiated thymidine. The method is described in detail by Basketter et al.[8,9] It is possible to use LLNA dose response studies to determine the dose required to produce a defined degree of sensitisation (typically the EC3 value) by interpolating the dose response plot.[1,10,11]

Patlewicz et al developed SARs for classifying the skin sensitizing potential of 17 aldehydes as strong, moderate or weak skin sensitisers.[12] The aldehydes were grouped into four distinct sub-categories of functionally related aldehydes: aryl-substituted aliphatic, aryl, aryl with special features (that can undergo metabolism) and α,β-unsaturated aldehydes. When dealing with aldehydes as a generic class of compounds, little correlation was found between sensitisation potential and generic chemical properties related to the aldehyde group alone. However good correlations were observed using the RAI model within the α,β-unsaturated aldehydes sub-category. Chemical reactivity was modelled using the σ^* value of the R (alkyl group). The relative sensitisation potency was defined as:

$$0.99\sigma^* +1.5$$

A good linear correlation was observed with a r^2 of 0.998. The equation of the straight line was:

$$y = 0.09856x + 1.4963$$

where x is $0.99\sigma^* + 1.5$ and y is log (1/EC3)*

where y = -log (EC3) + log (molecular weight).

Despite the complexities and still limited understanding of some of the processes leading to skin sensitization, it is possible to describe some of the relationships between chemical structures and the ability to form covalent conjugates with proteins. This knowledge which relates chemical structure to a specific endpoint can be programmed into expert systems. A toxicity prediction expert system is a computational program that embodies a range of QSARs or other knowledge that can be used to predict the toxicity of chemicals, including those which have not been synthesised and for which no safety data are available. DEREK (Deductive Estimation of Risk from Existing Knowledge) (http://www.chem.leeds.ac.uk/LUK/derek/index.html) is a knowledge-based expert system which reflects the current state of knowledge of structuretoxicity relationships with an emphasis on the understanding of mechanisms of toxicity and metabolism. The knowledge base covers a wide variety of important toxicological end-points, which include carcinogenicity, mutagenicity, skin sensitisation, teratogenicity, irritancy, and respiratory sensitisation. The expert knowledge incorporated into the DEREK system originated from Sanderson and Earnshaw.[13] These workers identified a series of substructures associated with certain types of toxic activity. These 'structural alerts' are codified in the rulebase so that when an 'unknown' structure is analysed by the software, the system essentially carries out a pattern recognition process to identify similar structural features. A strength of the DEREK expert system is that it can evolve as new knowledge is gained.[14] The latest version of the DEREK software incorporates a reasoning engine that uses log P and an algorithm for skin permeability in order to predict a chemical's propensity to induce skin sensitisation in humans.

Table 1. *Typical chemistries that can result in sensitisation involve reactions between xenobiotic electrophiles* and protein nucleophiles that lead to the formation of a covalent bond between the electrophilic compound and protein resulting in a protein-hapten complex****

i) Molecules with a polarized bond

δ^+ δ^- —CH_2—X

δ^+ δ^- C=O

δ indicated a partial charge

an arrow indicates the electrophilic centre susceptible to nucleophilic attack

X = Cl, Br or I

ii) Unsaturated compounds conjugated with electron withdrawing groups

H_2C=C(H)–CH(X)–R

NO_2

X = Cl, Br or I

iii) Typical protein nucleophiles

—$(CH_2)_4$—NH_2

lysine
(Lys or K)

—CH_2—**SH**

cysteine
(Cys or C)

iv) Reaction between DNCB electrophile and protein nucleophile

P—Nu: + Cl N O O O=N O → P Nu Cl O N O O=N O → P Nu O N O O=N O + Cl^-

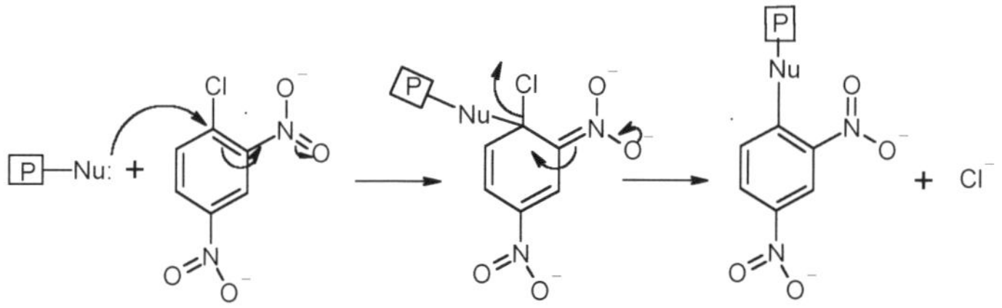

*as described in i and ii; **as shown in iii; **the reaction between DNCB and protein results in a protein-dinitrophenyl (DNP) complex, where DNP is the hapten.

Table 1 summarises the reactive chemistry normally associated with the potential to cause skin sensitisation.

Table 2. The following classes of enzymes have all been detected in skin (see Smith & Hotchkiss, 2001). Examples of the metabolic reactions they perform are listed below.

Phase I Enzymes—Some Examples of Activation/Functionalisation Mechanisms		
Microsomal mixed function oxidase i.e., Cytochrome P450 (many isoforms) with NAD(P)H dehydrogenase	–	hydroxylation alcohol oxidation epoxidation N-, O- and S-dealkylation oxidative deamination N- and S-oxidation dehalogenation reduction of azo and nitro compounds ring cleavage of heterocyclic ring systems
Alcohol dehydrogenases	–	interconversion of alcohols and aldehydes
Aldehyde dehydrogenases	–	conversion of aldehydes to acids
Flavin-containing monooxygenases	–	oxidation of secondary and tertiary amines oxidation of imines and arylamines
Esterases	–	ester hydrolysis to yield acid and alcohol
Amidases	–	amide hydrolysis
Phase II Enzymes and Conjugation Reactions		
UDP-glucuronosyltransferase	–	glucuronidation at -OH, -COOH, NH_2 and -SH
Sulphotransferase	–	sulphation at -NH_2, -SO_2NH_2 and OH
Methyltransferase	–	methylation at -OH and -NH_2
Acetyltransferase	–	acetylation at -NH_2, -SO_2NH_2 and -OH
Glutathione S-transferase	–	glutathione conjugation at epoxides and halides
Miscellaneous reactions	–	amino acid conjugation at -COOH condensation

Skin Metabolism

It is generally accepted that small molecules require covalent binding to macromolecular proteins to become immunogens. However, many small molecular compounds are chemically inert and are therefore not able to bind to proteins directly. Such compounds are termed prohaptens, which although nonelectrophilic in nature, may be converted into an electrophilic species. Prohaptens may be converted chemically—e.g., some chemicals, such as turpentine, may be oxidised to an electrophilic protein-reactive species upon exposure to air over time. Alternatively, prohaptens may be metabolised to more reactive species by xenobiotic metabolising enzymes. The skin has been recognised as an important site of extrahepatic metabolism. This is especially true for the epidermis, which constitutes part of the major interface between the body and the environment.

Xenobiotic metabolism is divided into at least two phases. Phase 1 reactions are predominantly mediated by cytochrome P450 isoenzymes,[15] although other Phase 1 enzymes e.g., alcohol and aldehyde dehydrogenases etc. also play roles in activating chemicals in the skin.[16] Metabolising enzymes likely to be involved in the generation of reactive species in skin are summarised in Table 2. Various conjugating enzymes e.g., sulphatases, glucuronidases,

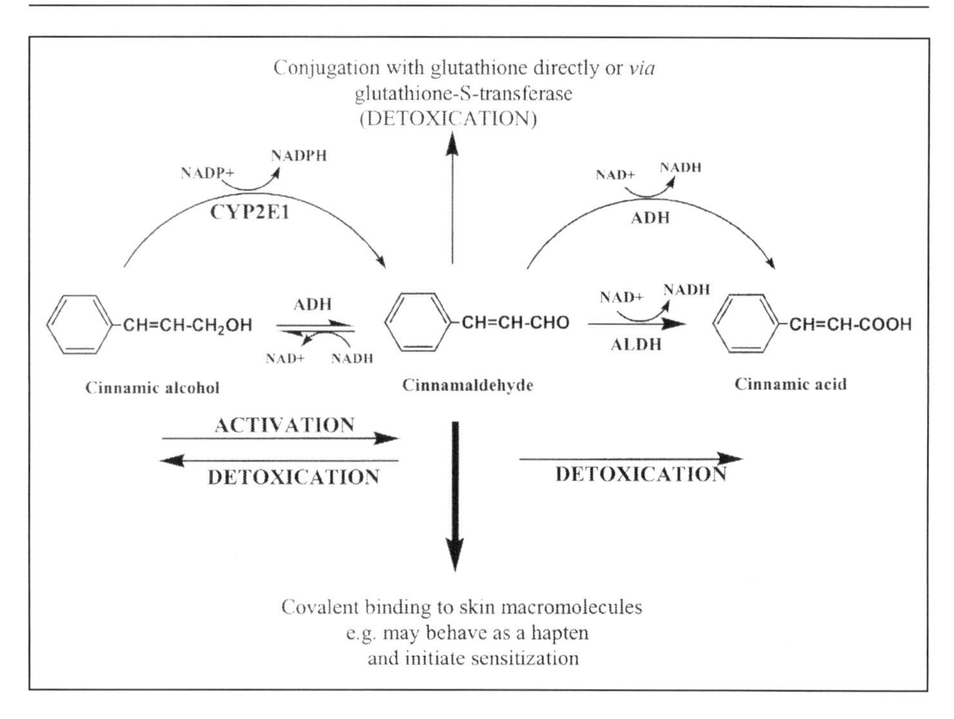

Figure 2. Proposed metabolism of cinnamaldehyde and cinnamic alcohol. Cinnamaldehyde (formula weight, FW = 132.16 g/mol) is an α, β-unsaturated aromatic aldehyde. Cinnamic alcohol (FW = 134.16 g/mol) is an aromatic alcohol and is the reduced form of cinnamaldehyde. The interconversion of cinnamaldehyde and cinnamic alcohol can be catalysed by alcohol dehydrogenase (ADH). Cinnamaldehyde can be irreversibly oxidised to cinnamic acid (FW = 148.16 g/mol) by either aldehyde dehydrogenase (ALDH), or ADH acting as a dismutase with NAD^+ as cofactor. Cinnamaldehyde may also bind to skin proteins or glutathione.

glutathione S-transferases etc. mediate phase 2 reactions, which normally lead to the formation of more hydrophilic derivatives that can be better eliminated from the body.[15] In general, such metabolism results in the detoxification of xenobiotics. However, if a highly reactive intermediate is formed that cannot be easily detoxified, it may reside in the skin for long enough to bind to nucleic acids or proteins, and could result in gene mutations in the former (leading to carcinogenesis) or hapten formation (and possibly sensitisation) with the latter. Hence, there is a balance between the phase 1 activation mechanisms and phase 2 detoxification mechanisms in the skin that can mediate toxicity.

A good example of the role skin metabolism may play is the potential activation of cinnamic alcohol to the presumed in vivo allergen, cinnamic aldehyde (Fig. 2).[17] Cinnamic alcohol is a chemically unreactive constituent of the European Standard Test Allergen 'Fragrance Mix' yet it is known to be an allergen. The action of alcohol dehydrogenase in the epidermis yielding the strongly sensitising cinnamic aldehyde would provide a potential explanation as to how cinnamic alcohol could give rise to ACD. In practice, it has proven hard to detect the formation of free (not protein-bound) cinnamic aldehyde from the corresponding alcohol in in vitro experiments using intact skin and skin homogenates.[18,19] However, the presence of protein-bound cinnamaldehyde has been observed in skin treated with cinnamic alcohol using cinnamaldehyde-specific immunodetection methods.[20] This suggests that any cinnamaldehyde formed within the skin is either detoxified rapidly (to cinnamic acid) or becomes protein bound. However, the picture may be more complex than is predicted. In the clinic, whilst there are a proportion of those patch test positive to cinnamic alcohol who also react with cinnamaldehyde (supporting the hypothesis of a common cinnamaldehyde-derived hapten), there are also a proportion who are only cinnamic alcohol positive and do not react to cinnamaldehyde. This clinical observation suggests that there may also be other mechanisms that could contribute to cinnamic alcohol allergy.[21] A true cross-reactivity patch test study has not been performed to date for these compounds.

Although our understanding of the role of metabolic activation (and inactivation) in ACD is in its infancy, it seems very probable that it is one of the key elements responsible for the inter-individual differences of human responses to chemical contact allergens and may provide the rationale for why some individuals tolerate a lifetime of exposure to allergens such as PPD, whilst others experience ACD after only modest amounts of exposure.

Through the recognition that the toxic properties of chemicals are inherent in their molecular structure, a body of knowledge has been developed with respect to the physicochemical parameters influencing skin penetration and the chemical substituents and 'structural alerts' that define chemical reactivity with proteins. This knowledge, particularly as embodied in a computer based expert system such as DEREK, can be employed to predict which chemicals are most likely to have the potential to cause skin sensitisation. Although the predictions are not yet wholly accurate, with the consequence that in vivo models are still the most reliable, progress continues to be made in this area. However, before proceeding to discuss the latest developments in that area, it is appropriate first to consider potential cell based in vitro systems.

In Vitro Predictive Models of Contact Sensitisation

The induction of cutaneous immune responses to chemical allergens requires the coordinated interaction between various cell types under the control of a myriad of molecular mediators. Such complexity leads to the inevitable conclusion that the development in vitro of cell or tissue culture models that recapitulate wholly or in part the initiation of skin sensitisation presents a not inconsiderable challenge. Approaches, or rather potential approaches, to skin sensitisation testing have recently been reviewed in detail elsewhere.[22-24] For the purposes of this article it is sufficient to identify only in general terms the strategies that have attracted some interest; these being based on responses induced in dendritic cells (DC) or in T lymphocytes.

DCs are a diverse family of bone marrow-derived cells that play pivotal roles in the initiation of adaptive immune responses. LCs form part of this family. They reside in the epidermis where they are though to act as sentinels for the immune system. It is now appreciated that in responses to topical sensitisation LCs become mobilized and are induced to migrate from the epidermis, via afferent lymphatics, to skin draining lymph nodes. A proportion of the cells that traffic from the skin carry high levels of antigen. En route from the skin, LCs are subject to a functional maturation such that by the time of their arrival in regional lymph nodes they have acquired the characteristics of mature immunostimulatory DC that able to present antigen effectively to responsive T lymphocytes. The immunobiological processes of migration and maturation are regulated by epidermal cytokines.[25-27] As LCs play such essential roles in the acquisition of skin sensitisation there has been considerable interest in the possibility that they exhibit characteristic responses following exposure to contact allergens, and that such responses might be modelled in vitro as the basis for an alternative approach to hazard identification. This thesis is given some credence by the observations made a decade ago that exposure of mice to contact allergens is associated with increases in cytokine expression by epidermal cells. Importantly some cytokines, such as interleukin 1β (IL-1β) were shown to be upregulated only by chemicals known to be skin sensitisers; nonsensitising skin irritants were without effect on IL-1β expression. This cytokine attracted further interest because, in mouse epidermis at least, it is a product only of LC.[28] A major constraint in capitalising on these observations was the fact that LC represent only a minority population in the epidermis and are difficult to isolate without compromising their functional integrity. The solution came with the first description of methods for growing DC with various phenotypic characteristics from relevant precursor cells drawn from either peripheral blood or bone marrow. One of the first approaches was described by Sallusto and Lanzavecchia.[29] These new methods provided the opportunity to question whether cultured LC-like cells would display an increase in IL-1β mRNA expression following exposure to contact allergens. The first experiments were encouraging,[30] and it does appear that contact allergens (or at least potent contact allergens) are able to induce increased IL-1β expression by cultured DC in some circumstances.[30-32] However, there are various features of such responses that suggest this will not, at the present time, provide a suitable basis for the development of a robust in vitro method. Importantly, even with the most potent skin sensitising chemicals, only modest increases in the levels of expression of mRNA for IL-1β are observed; possibly because the gene that encodes this cytokine has only a limited dynamic range. Moreover, responses are somewhat variable; that is donor-dependent differences in responses have been observed with DC prepared from some the blood of some individuals failing consistently to respond to allergen treatment with increased IL-1β expression.[31,32] Although IL-1β may not provide a sufficiently sensitive readout for the interaction of DC with chemical allergens this does not invalidate the strategy altogether. It is clear that during the induction phase of skin sensitisation LC are subject to a variety of important changes, including the altered expression of cytokines, of chemokine and cytokine receptors, of other membrane determinants and of intracellular signalling pathways. It is relevant therefore to ask whether there are other allergen-induced changes provoked in LC-like cells by contact allergens that would provide a sensitive and selective marker of skin sensitising activity. One way in which to address this question is to characterise more holistically changes in gene expression associated with exposure to skin sensitising chemicals. To this end, an attractive strategy is the application of microarray transcript profiling methods,[33] and such experiments are currently in progress in a number of laboratories across the world.

The other main strategy that has been considered is the evaluation in vitro of T lymphocyte responses induced by chemical allergens. Although it is relatively straightforward to provoke antigen-induced secondary responses in vitro using T lymphocytes, what is far more of a challenge is to stimulate antigen-specific primary activation using as responder cells T lymphocytes from naïve donors. Nonetheless, some progress has been made and it has been reported that proliferative responses by naïve T lymphocytes can be induced in culture with some, but not all, chemical allergens tested. See, for example, Rougier et al[34] It remains to be seen how useful a strategy this is for development of a robust in vitro approach to sensitisation testing.

LLNA: Hazard Identification

Hazard identification represents the primary step in the safety evaluation of a chemical substance and so addresses the question of whether the chemical possesses, intrinsically, the propensity to cause contact allergy. The LLNA represents a new option in this area, having recently been specifically validated for this purpose.[35] The adoption of a new method into regulatory guidelines represents a substantial challenge, demanding both scientific consensus on its suitability as well as acceptance via the formal processes prescribed for validation. This step was taken for the LLNA via the US Inter-agency Coordinating Committee on the Validation of Alternative Methods (ICCVAM). This independent review has been published;[36] ICCVAM concluded that the LLNA was fully valid as a standalone alternative to existing guinea pig tests. Following this, the LLNA was adopted by agencies in the USA as an accepted skin sensitisation method (e.g., EPA, FDA, OSHA). Additionally, the LLNA has been prepared as a new Test Guideline (No 429) by the Organisation for Economic Cooperation and Development (OECD).[37] Following approval by National Representatives in 2001, publication of Test Guideline 429 in April 2002, available on the OECD website (http://www.oecd.org).

In parallel, the European Union (EU) has prepared a new test method on the LLNA (B42);[38] the text closely follows that prepared by the OECD.

LLNA: Risk Assessment

Although the LLNA provides a simple means for the predictive identification of chemical with the potential to behave as skin allergens, this in itself is insufficient to enable a safety assessment to be made. The risks presented to humans from exposure to allergenic chemicals depend on several variables related to exposure (frequency and duration of contact, the exposure concentration etc) and also on the potency of the allergen. Whilst the exposure variables do not depend on the chemical itself (but rather how it is used), allergen potency is a function of the chemical, and this can be measured by the LLNA. Indeed, the use of LLNA data to estimate the relative potency of an allergen is a topic which has received substantial consideration since its first proposal a decade ago.[39] Important points to consider are:

- How is potency of an allergen estimated in the LLNA?
- The robustness of this potency estimation
- Relevance of the potency estimation to humans
- How to apply potency measures in risk assessment

In the LLNA, potency is estimated by derivation of the EC3 value, the concentration of a test chemical necessary to produce a 3-fold stimulation of proliferation in the draining lymph nodes compared to concurrent vehicle treated controls. The logistics underpinning this have been published.[40,41] The second of these considerations, robustness, simply means that the measurement of potency is reproducible, both within and between laboratories over time. That this is the case has been fully demonstrated.[42–45] Perhaps the most important of the points is the demonstration that EC3 values derived from the LLNA correlate with what is understood about the relative potency of skin allergens in humans. Several publications have already indicated that this is indeed the case[46–48] and the currently available data is summarised in Table 3.

Ultimately to be of practical value, potency information has to be capable of being factored into a risk assessment. Generic approaches to skin allergy risk assessment have already been published,[49] but more recently, the manner in which EC3 values could be incorporated, providing a more definitive and quantitative view on allergen potency, has also been described.[50,51] In principle, the EC3 value provides an objective route to the estimation of no effect levels for humans. Such information in combination with exposure considerations can then form the core of a quantitative risk assessment approach for contact allergy.[52–54] For regulatory purposes, the EC3 value might provide a simple route through which to grade allergens as strong, moderate and weak;[52] in the EU, this might then permit the establishment of varying default threshold limits for contact allergens, rather than having the standard current value of 1%. For example, chemicals classified as strong allergens, could have a default limit 10 or even 100 times lower, whilst the weakest allergens could have a default limit above 1%. In this way, risk management could be effectively targeted towards chemicals that presented the greater risk to human health.

Little mention has been made of the more traditional guinea pig methods for the identification and assessment of skin sensitisation hazards in this chapter—this is deliberate, not least since these methods are much better suited to simple hazard identification as opposed to the measurement of relative potency.[55,56]

Concluding Remarks

A body of knowledge on the physicochemical parameters governing skin penetration and the chemical substructures associated with ACD, embodied in the computer based expert system DEREK, can help to predict which chemicals are most likely to have the potential to cause skin sensitisation. However, to make meaningful risk assessments, the potency of identified

Table 3. LLNA potency estimates compared to human classification

Chemical	LLNA[a] EC3[b] Value (%)	Human Potency
Oxazolone	0.01	Extreme
Diphencyclopropenone	0.05	Extreme
Methyl/chloromethylisothiazolinone	0.05	Extreme
p-Phenylenediamine	0.06	Extreme
2,4-Dinitrochlorobenzene	0.08	Extreme
Toluene diisocyanate[c]	0.11	Strong
Glutaraldehyde	0.20	Strong
Trimellitic anhydride[c]	0.22	Strong
Phthalic anhydride[c]	0.36	Strong
Formaldehyde	0.40	Strong
Methylisothiazolinone	0.40	Strong
Isoeugenol	1.3	Strong
Cinnamic aldehyde (cinnamal)	2.0	Strong
Diethylmaleate	2.1	Strong
Zinc dimethyldithiocarbamate[c]	2.7	Strong
Phenylacetaldehyde	4.7	Strong
Methyldibromo glutaronitrile	5.2	Strong
Dipentamethylenethiuramdisulphide[c]	5.2	Strong
Tetramethylthiuramdisulfide	6.0	Strong
4-Chloroaniline	6.5	Strong
Hexylcinnamic aldehyde (hexyl cinnamal)	8.0	Moderate
Mercaptobenzothiazole	9.7	Moderate
Abietic acid	11	Moderate
Citral	13	Moderate
Eugenol	13	Moderate
p-Methylhydrocinnamic aldehyde	14	Moderate
Mercaptobenzimidazole[c]	15	Moderate
p-tert-Butyl-a-methyl hydrocinnamal	19	Moderate
Hydroxycitronellal	20	Moderate
Cyclamen aldehyde	21	Moderate
Dipentamethylenethiuramtetrasulphide[c]	21	Moderate
Benzocaine[c]	22	Moderate
5-Methyl-2,3-hexanedione	26	Moderate
Linalool	30	Moderate
Ethyleneglycol dimethacrylate	35	Weak
Diethanolamine[c]	40	Weak
Isopropyl myristate	44	Weak
Propyl paraben	>50	Weak
Vanillin	>50	Weak
Aniline	>50	Weak
Ethyl vanillin	>50	Weak
Acetanisole	Non-sensitizing	Non-sensitizing
Dextran	Non-sensitizing	Non-sensitizing
Diethylphthalate	Non-sensitizing	Non-sensitizing
Glycerol	Non-sensitizing	Non-sensitizing
Hexane	Non-sensitizing	Non-sensitizing
Isopropanol	Non-sensitizing	Non-sensitizing
Octanoic acid	Non-sensitizing	Non-sensitizing
Resorcinol	Non-sensitizing	Non-sensitizing
Tween 80	Non-sensitizing	Non-sensitizing

[a] LLNA, local lymph node assay; [b] EC3 value is the estimated concentration of a chemical necessary to give a 3 fold stimulation of proliferation in lymph nodes compared to concurrent vehicle-treated controls. The vehicle for these studies was actone:olive oil, 4:1, v/v.; [c] Data for a small number of chemicals involved a modified version of the LLNA not performed entirely in accord with OECD 429[37] and is therefore annotated.

allergens must be measured. This can now be achieved using the dose response data from the LLNA. Combination of this information with a thorough understanding of likely skin exposure to the allergen permits a proper risk assessment to be undertaken and appropriate risk management measures can be implemented. The challenge facing toxicologists working in this area is how to achieve this goal in the future by the sole use of in vitro and in silico methods. Key to this will be the continuing research on the underlying mechanisms of skin sensitisation and in particular the identification of the determinants of allergen potency.

References

1. Lepoittevin JP, Basketter DA, Goossens A et al, eds. Allergic Contact dermatitis: The molecular basis. Berlin: Springer-Verla, 1997.
2. Flynn GL. Physicochemical determinants of skin absorption. In: Gerrity TR, Henry CL, eds. Principles of Route-to-Route Extrapolation for Risk Assessment. New York: Elsevier, 1990:93-127.
3. Dupuis and Benezra. Contact Dermatitis to Simple Chemicals: A Molecular Approach. New York: Marcel Dekker, 1982.
4. Basketter D, Dooms-Goossens A, Karlberg AT et al. The chemistry of contact allergy: Why is a molecule allergenic? Contact Dermatitis 1995; 32:65-73.
5. Roberts DW, Williams DL. The derivation of quantitative correlations between skin sensitisation and physicochemical parameters for alkylating agents and their application to experimental data for sultones. J Theor Biol 1982; 99:807-825.
6. Barratt MD, Basketter DA, Roberts DW. Quantitative structure activity relationships. In: LePoittevin J-P, Basketter DA, Dooms-Goossens A et al, eds. The Molecular Basis of Allergic Contact Dermatitis. Heidelberg: Springer-Verlag, 1997:129-154.
7. Enslein K, Gombar VK, Blake BW et al. A QSTR Model for the Dermal Sensitisation Guinea Pig Maximization Assay. Food Chem Toxicol 1997; 35:1091-1098.
8. Kimber I, Basketter DA. The murine local lymph node assay; collaborative studies and new directions: A commentary. Food Chem Toxicol 1992; 30:165-169.
9. Basketter DA, Blaikie L, Dearman RJ et al. Use of the local lymph node assay for the estimation of relative contact allergenic potency. Contact Dermatitis 2000; 42:344-348.
10. Basketter DA, Roberts DW, Cronin M et al. The value of the LLNA in quantitative structure activity investigations. Contact Dermatitis 1992; 27:137-142.
11. Roberts DW, Basketter DA. QSARs: Sulfonate esters in the LLNA. Contact Dermatitis 2000; 42:154-161.
12. Patlewicz G, Basketter DA, Smith CK et al. Skin sensitisation structure activity relationships for aldehydes. Contact Dermatitis 2001; 44:331-336.
13. Sanderson DM, Earnshaw CG. Computer prediction of possible toxic action from chemical structure; The DEREK system. Human Exp Toxicol 1991; 10:261-273.
14. Barratt MD, Langowski JJ. Validation and Subsequent Development of the DEREK Skin Sensitization Rulebase by Analysis of the BgVV List of Contact Allergens. J Chem Inf Comput Sci 1999; 39:294-298.
15. Smith CK, Hotchkiss SAM. Allergic Contact Dermatitis: Chemical and metabolic mechanisms. London: Taylor & Francis Ltd., 2001.
16. Cheung C, Smith CK, Hoog J-O et al. Expression and localisation of human alcohol and aldehyde dehydrogenase enzymes in skin. Biochem Biophys Res Commun 1999; 261:100-107.
17. Basketter DA. Skin sensitization to cinnamic alcohol: The role of skin metabolism. Acta Derm Venereol 1992; 72:264-265.
18. Moore CA, Smart ATS, Hotchkiss SAM. Human skin absorption and metabolism of the contact allergens, cinnamic aldehyde and cinnamic alcohol. Toxicol Appl Pharmacol 2000; 168:189-199.
19. Cheung C, Hotchkiss SA, Pease CK. Cinnamic compound metabolism in human skin and the role metabolism may play in determining relative sensitization potency. J Dermatol Sci 2003; 31:9-19.
20. Elahi EN, Wright Z, Hinselwood D et al. Protein binding and metabolism influence the relative skin sensitization potential of cinnamic compounds. Chem Res Toxicol 2004; 17:301-310.
21. Personal communication with St. John's Institute of Dermatology, London UK.

22. Kimber I, Pichowski JS, Basketter DA et al. Immune responses to contact allergens: Novel approaches to hazard evaluation. Toxicology Letters 1999; 106:237-246.
23. Kimber I, Pichowski JS, Betts CJ et al. Alternative approaches to the identification and characterization of chemical allergens. Toxicol in Vitro 2001; 15:307-312.
24. Ryan CA, Hulette BC, Gerberick GF. Approaches for the development of cell-based in vitro methods for contact sensitization. Toxicol in Vitro 2001; 15:43-55.
25. Kimber I, Dearman RJ, Cumberbatch M et al. Langerhans cells and chemical allergy. Curr Opin Immunol 1998; 10:614-619.
26. Kimber I, Cumberbatch M, Dearman RJ et al. Cytokines and chemokines in the initiation and regulation of epidermal Langerhans cell mobilization. Brit J Dermatol 2000; 142:401-412.
27. Cumberbatch M, Dearman RJ, Griffiths CEM et al. Langerhans cell migration. Clin Exp Dermatol 2000; 25:413-418.
28. Enk AH, Katz SI. Early molecular events in the induction phase of contact sensitivity. Proc Nat Acad Sci USA 1992; 89:1398-1402.
29. Sallusto F, Lanzavecchia A. Efficient presentation of soluble antigen by cultured human dendritic cells is maintained by granulocyte/macrophage colony stimulating factor plus interleukin-4 and down-regulated by tumor necrosis factor α. J Exp Med 1994; 179:1109-1118.
30. Reutter K, Jager D, Degwert J et al. In vitro model for contact sensitization: II Induction of IL-1β mRNA in human blood-derived dendritic cells by contact sensitizers. Toxicol in Vitro 1997; 11:619-626.
31. Pichowski JS, Cumberbatch M, Dearman RJ et al. Investigation of induced changes in interleukin 1β mRNA expression by cultured human dendritic cells as an in vitro approach to skin sensitization testing. Toxicol in Vitro 2000; 14:351-360.
32. Pichowski JS, Cumberbatch M, Dearman RJ et al. Allergen-induced changes in interleukin 1β (IL-1β) mRNA expression by human blood-derived dendritic cells: Inter-individual differences and relevance for sensitization testing. J Appl Toxicol 2001; 21:115-121.
33. Pennie WD, Kimber I. Toxicogenomics; transcript profiling and potential application to chemical allergy. Toxicol in vitro 2002; 16:319-326.
34. Rougier N, Redziniak G, Schmitt D et al. Evaluation of the capacity of dendritic cells derived from blood $CD34^+$ precursors to present haptens to unsensitized autologous T cells in vitro. J Invest Dermatol 1998; 110:348-352.
35. Gerberick GF, Ryan CA, Kimber I et al. Local lymph node assay validation assessment for regulatory purposes. Am J Cont Derm 2000; 11:3-18.
36. NIH 1999. The murine local lymph node assay: a test method for assessing the allergic contact dermatitis potential of chemicals/compounds. NIH No. 99-4494.
37. OECD 2002. Guidelines for Testing of Chemicals. Guideline No. 429. Skin Sensitisation: The Local Lymph Node Assay. Organisation for Economic Cooperation and Development, Paris.
38. UK HSE: Formal statement on the acceptability of the LLNA
39. Kimber I, Basketter DA. Contact sensitization: A new approach to risk assessment. Human and Ecological Risk Assessment 1997; 3:385-395.
40. Basketter DA, Lea L, Cooper K et al. Thresholds for classification as a skin sensitiser in the local lymph node assay: a statistical evaluation. Food Chem Toxicol 1999; 37:1167-1174.
41. Basketter DA, Lea L, Cooper K et al. A comparison of statistical approaches to derivation of EC3 values from local lymph node assay dose responses. J Appl Toxicol 1999; 19:261-266.
42. Loveless SE, Ladics GS, Gerberick GF et al. Further evaluation of the local lymph node assay in the final phase of an international collaborative trial. Toxicol 1996; 108:141-152.
43. Dearman RJ, Hilton J, Evans P et al. Temporal stability of local lymph node assay responses to hexyl cinnamic aldehyde. J Appl Toxicol 1998; 18:281-284.
44. Warbrick EV, Dearman, RJ, Lea LJ et al. Local lymph node assay responses to paraphenylenediamine: intra- and inter-laboratory evaluations. J Appl Toxicol 1999; 19:255-260.
45. Dearman RJ, Wright ZM, Basketter DA et al. The suitability of hexyl cinnamic aldehyde as a calibrant for the using local lymph node assay. Contact Dermatitis 2001; 44:357-361.
46. Basketter DA, Blaikie L, Dearman RJ et al. Use of the local lymph node assay for the estimation of relative contact allergenic potency. Contact Dermatitis 2000; 42:344-348.

47. Ryan CA, Gerberick GF, Cruse LW et al. Activity of human contact allergens in the murine local lymph node assay. Contact Dermatitis 2000; 43:95-102.
48. Gerberick GF, Robinson MK, Ryan CA et al. Contact allergenic potency: Correlation of human and local lymph node assay data. Am J Cont Derm 2001; 12:156-161.
49. Robinson MK, Gerberick GF, Ryan CA et al. The importance of exposure estimation in the assessment of skin sensitization risk. Contact Dermatitis 2000; 42:251-259.
50. Gerberick GF, Robinson MK, Felter S et al. Understanding fragrance allergy using an exposure-based risk assessment approach. Contact Dermatitis 2001; 45:333-340.
51. Basketter DA, Gerberick GF, Kimber I. Measurement of allergenic potency using the local lymph node assay. Trends Pharm Sci 2001; 22:264-265.
52. Basketter DA, Evans PE, Gerberick GF et al. Factors affecting thresholds in allergic contact dermatitis: Safety and regulatory considerations. Contact Dermatitis 2002:47,1-6.
53. Felter SP, Ryan CA, Basketter DA et al. Application of the risk assessment paradigm to the induction of allergic contact dermatitis. Reg Toxicol Pharmcol 2002; accepted.
54. Felter SP, Robinson MK, Basketter DA et al. A review of the scientific basis for default uncertainty factors for use in quantitative risk assessment of the induction of allergic contact dermatitis. Contact Dermatitis 2002; accepted.
55. Kimber I, Basketter DA, Berthold K et al. Skin sensitisation testing in potency and risk assessment. Toxicol Sci, 2001:59:198-208.
56. Basketter DA, Kimber I. Predictive testing in contact allergy: facts and future. Allergy 2001; 56:937-943.

Chapter 2

Molecular Recognition of Haptens by T Cells:

More Than One Way to Tickle the Receptor

Hans Ulrich Weltzien, Andrea Dötze, Katharina Gamerdinger, Sven Hellwig and Hermann-Josef Thierse

Summary

Haptens as low molecular chemicals compose a major percentage of the universe of allergens, particularly with respect to allergic contact dermatitis (ACD). They are usually defined as compounds which only upon covalent interaction with proteins acquire the potential to induce hapten-specific B cell as well as T cell responses. Here we discuss recent developments in the basic understanding of MHC-restricted recognition of haptens by αβ T cells. Major emphasis is put on the structural elucidation of MHC-restricted epitopes for the model hapten trinitrophenyl (TNP) as well as for nickel as a major human contact sensitizer. Furthermore, we show that in addition to forming allergenic determinants penicillin and nickel can interact with the immunologically relevant molecules interferon gamma (IFN-γ) and heatshock proteins, respectively. Also taking into account that coordination complexes of metal ions as well as certain drugs may be recognized by T cells in noncovalent MHC association, the definition and functions of haptens may have to be reevaluated.

Introduction

T cell receptors for antigen (TCR) of αβ T cells, unlike B cell receptors (BCR), recognize antigenic peptides presented in the binding grooves of self-MHC-molecules.[1] However, parallel to the first description of MHC-restricted recognition of viral proteins,[2] Shearer already reported that T cells also revealed MHC-restricted reactivity towards a nonpeptide chemical hapten, i.e., the lysine-reactive trinitrobenzene sulfonic acid (TNBS).[3] It was obvious that trinitrophenyl (TNP) epitopes for T cells as those for B cells required covalent binding of the hapten to protein carriers. Nevertheless, molecular explanations for MHC-restricted recognition of haptens as well as of proteins only became possible upon the discovery of the TCR[4] and after Grey and colleagues had identified MHC-associated peptides as the antigenic epitopes for T cells.[5] More recently this understanding resulted in a remarkably detailed description of MHC-peptide-TCR contacts[6-8] as well as of the biochemical signals induced by their formation.[9] These studies had a wide-ranging impact on allergy research since it soon became evident that T cells were central players not only in delayed type hypersensitivities, but also in IgE mediated Type I allergies.[10-12] For allergic contact dermatitis (ACD) it was apparent that most

Immune Mechanisms in Allergic Contact Dermatitis, edited by Andrea Cavani and Giampiero Girolomoni. ©2005 Eurekah.com.

contact allergens, such as chemicals, drugs, plant ingredients or metals were clearly of nonpeptide nature. This may explain why hapten-specific T cells have received increasing attention among allergologists even though basic immunology has turned its interest to other topics. The latter phenomenon is somehow surprising since T cell activation by haptens is far from being fundamentally understood, and on the other hand our modern environment constantly supplies us with thousands of potentially T cell reactive man-made or natural haptens.

Hapten Epitopes for T Cells

MHC-Associated Hapten-Peptide Conjugates

Following Shearer's report on TNP-specific T cells,[3] T lymphocytes with specificities for numerous other protein-reactive chemicals were described.[13] It became apparent that T cells, like immunoglobulins, were capable of responding to almost any chemical structure, provided it could covalently modify proteins. Early experimental evidence suggested that hapten epitopes for T cells did not necessarily require covalent modification of MHC proteins,[14] but rather might originate from haptenization of non-MHC molecules.[15,16] About 10 years ago it was Ortman et al[17] for TNP and Nalefski and Rao[18] for p-azobenzarsonate who first defined hapten-modified, MHC-binding peptides as the major antigenic epitopes for hapten-specific T cells.

For TNP and DNP these epitopes, as well as the corresponding TCR, were studied for a variety of murine class I and class II MHC-restricted T cells.[19] These data revealed that TNP-modified, MHC-bound peptides, indeed, represented the major type of epitopes recognized by in vitro TNBS-induced $CD8^+$ as well as $CD4^+$ T cells. It was further found that the majority of TNP-specific T cells were rather indifferent concerning the amino acid sequences of the modified, MHC anchoring carrier peptides, but highly specific for the position of the TNP-carrying lysine within this sequence.[20-22] Notably, dominant TNP-epitopes were produced when TNP-Lys replaced amino acids which also represented a central antigenic position in unmodified peptides, e.g., position 4 in H-$2K^b$ binding octapeptides.[23] The relative carrier-indifference but position-specificity as well as rapidly increasing knowledge concerning peptide binding motifs for individual MHC alleles[24] allowed for chemical synthesis of MHC-allele-specific hapten-peptide conjugates. This, in turn, accomplished allele-specific hapten modification of MHC molecules on living cells.

The relevance of hapten-peptide epitopes in ACD was proven most convincingly by sensitization of mice with TNP-peptides for subsequent elicitation of ear swelling reactions by TNCB solution.[25] Injection of MHC class I or class II binding TNP-peptides revealed that depending on the route and mode of injection mice were sensitized for TNCB either by $CD4^+$ or $CD8^+$ T cells.[26] Subcutaneous injection in incomplete Freund's adjuvant favoured class II binding peptides whereas injection of peptide-treated dendritic cells (DCs) worked better with class I binding TNP-peptides. This latter finding compares well with recent data of Martin et al demonstrating that skin painting of mice with TNCB or injection of TNBS-modified DCs selectively activates $CD8^+$ cells although MHC I deficient animals mounted strong CD4 responses.[27,28] Activation of penicillin-specific T cells from allergic donors by synthetic, HLA-binding penicilloyl-peptides[29] revealed a relevance of hapten-peptides also in human allergy.

Generation of Hapten-Modified, MHC Binding Peptides in Living Cells

Synthetic hapten-peptides with appropriate MHC binding motifs were shown in vitro to bind effectively to MHC molecules on living cells.[30,31] MHC-associated TNP-peptides have also been recovered from TNBS treated cells.[20] However, in the case of TNBS modification of living cells there are several options to create MHC-bound hapten-peptides. Firstly, treatment of isolated H-$2K^b$ molecules with TNBS revealed direct modification of MHC-associated self-peptides.[32] Secondly, endocytosis and processing of hapten-conjugated external or membrane proteins may generate intracellular hapten-peptides. Such peptides would be expected to be preferentially presented on MHC class II. However, cellular uptake of TNP-ovalbumin via acid labile liposomes also resulted in MHC class I restricted TNP presentation.[33] Finally, hydrophobic haptens might penetrate plasma membranes and directly modify intracellular proteins. Processed hapten-peptides might then be presented on either class I or class II MHC molecules. For chemically reactive haptens direct experimental evidence for this latter model is still missing.

This is different for those xenobiotics which per se are not chemically reactive. They have to enter the cell to be metabolized into reactive intermediates which then covalently modify intracellular proteins.[34,35] Such so-called pro-haptens encompass molecules like urushiol from poison ivy,[36] large numbers of drugs, e.g., sulfonamides or anaesthetics,[37] hair dyes such as p-phenylene-diamine[38] and many others,[39] in fact probably the majority of nonprotein allergens. Some reactive metabolites of the above mentioned compounds have been identified and shown in animal models as well as in in vitro experiments to result in MHC-associating hapten-peptides.

TCR Contacts to MHC-Peptide Associated Haptens

The molecular interaction of T cell antigen receptors with MHC-restricted hapten epitopes has been studied for several different haptens by various groups. For the hapten TNP investigators were faced with the intriguing observation that some nonimmunized mouse strains contained extraordinary high frequencies of hapten-reactive T cells.[40] Original reports of a superantigen-like preferential use of selected TCR V-elements in TNP recognition[41] were corrected later by detailed sequence analyses.[42] These data showed some preference for AV10 and JB2S6 usage in H-$2K^b$ restricted TNP-specific T cells, but in terms of repertoire size and selectivity of MHC restriction revealed no principal difference to nominal peptide responses. Some hapten-specific T cell clones exhibited extremely promiscuous, i.e., allele-independent, MHC restriction.[22,43,44] Such clones may play an important role in allergic recall responses. However, they are too rare to account for the high frequencies of hapten-reactive T cells.

The abundance of hapten-reactive TCR might be better explained by the fact that many cross-reactive hapten-epitopes are produced by position-specific modification of very different peptides.[25,30,31,33,45] In these cases TCR-peptide contacts contribute little to the overall TCR-antigen affinity which is dominated by direct TCR-hapten contacts. Moreover, the majority of TNP-specific TCR prefer the hapten attached in central positions of the MHC-associated peptides.[19] In this respect it is of interest that Wilson and coworkers[46] described a large central hydrophobic, potentially hapten-accomodating cavity in the antigen contact zone of TCR crystals. This suggests that large numbers of similar, but not identical TCR could eventually be activated by a whole panel of different hapten-peptides, provided an identical positioning of the hapten. Unfortunately, however, no MHC-hapten-TCR complex has so far been crystallized.

The most detailed information on TCR contacts with MHC-associated hapten-peptides has been obtained by the group of Luescher,[47,48] using photoaffinity labeled peptides and haptens. Light induced covalent crosslinking of contact sites in such TCR-MHC complexes revealed that hapten-peptides are contacted by TCR absolutely comparably to nominal peptide antigens. Thus, hapten-peptides may well be applied to study basic phenomena of TCR-antigen interactions. They have the additional advantage that they may be visualized and quantified by surface staining with hapten-specific antibodies which also may be applied to specifically block the TCR-MHC interaction. Thus, Preckel et al[31,49,50] have used TNP- and dinitrophenyl (DNP)-specific CTL clones to study the influence of ligand alterations on TCR activation. Alteration of the agonist ligands either in their peptide sequence (altered peptide ligands) or in the structure of their hapten component (altered hapten ligands) resulted in partial agonists or even antagonists, some of which were shown to induce a state of anergy in the specific CTL.

Metals As T Cell Antigens

Metals are astonishingly frequent inducers of a variety of T cell mediated human diseases. Chronic berryllium disease (CBD),[51] cobalt hard metal lung disease (HMD),[52] and contact hypersensitivities to a variety of metals, most notably to nickel,[53-55] are prominent examples. In all these cases the induction of metal-specific, MHC-restricted T cells is well established, but the nature of the antigenic epitopes and the mode of T cell activation by metals is still under debate. Unlike other haptens metal ions do not modify proteins by covalent interaction, but by means of reversible coordination complexes, usually involving four or six coordination sites in a metal-specific geometric arrangement. Ni, for instance, requires an octahedral arrangement of six ligands of which four form a square planar arrangement and the other two may also be occupied by water.[56]

Such complexes are difficult to identify on cell surfaces, and one therefore relies to a large extent on hypothetical models. Three such models have been proposed to explain the activation of metal-reactive T cells: (1) Metal-specific TCR react to determinants formed by complexation of metal ions with MHC embedded self peptides,[54,57] (2) TCR recognize metal-modified amino acid residues of the MHC molecule itself or metal-provoked conformational changes in the MHC,[58] and (3) metals affect the processing of self antigens, resulting in T cells reactive to cryptic self-peptides not containing metal ions as part of the antigenic epitope.[59]

None of these models is mutually exclusive, and each metal may use several pathways to activate T cells. The first and third of these models have been suggested for Ni- as well as for Au-reactive T cells.[54] The second model is favoured in the case of Co (HMD) and Be (CBD) where the development of disease has been correlated to HLA-DPB1* alleles expressing glutamate in position 69 of the DP β-chain.[58,60,61] Glu69 has been implicated in the binding of Be as well as Co. Finally, the third model has been explicitely proven for Au treated proteins, but is also likely to hold true for other metals.[62,63]

T Cell Recognition of Nickel

In recent years research on metal reactive T cells has clearly focussed on Ni, which not only causes the most frequent type of metal allergy, but also represents the most common contact allergen for humans.[55] Ni-reactive $CD4^+$ and $CD8^+$ T cell lines and clones have been isolated from peripheral blood as well as from skin lesions of sensitized individuals and their MHC-restriction, metal cross-reactivity, and activation requirements have been examined.[64-69]

No significant association of HLA alleles with Ni-induced ACD was observed in population studies.[70] However, in individual donors we found clear preferences, at least in the case of $CD4^+$ Ni-reactive T cells, for one of the patients HLA class II alleles (C. Moulon, unpublished observation). Moreover, stimulating $CD4^+$ clones with differently treated antigen presenting cells (APC) revealed several functionally distinguishable T cell populations. About half of the cells could be stimulated by Ni in the presence of glutaraldehyde fixed APC, the others required non fixed stimulators.[19,67] Both populations could be further subdivided based on the ability or inabilty to respond to APC that were pulsed with Ni and washed prior to the stimulation assay.

We take this to indicate different types of TCR reacting to principally different types of Ni-induced determinants. HLA-restricted TCR which react to fixed, Ni-pulsed APC probably recognize epitopes formed by stable complexes of Ni ions with MHC or/and MHC-associated peptides on the APC. In contrast, cells requiring metabolically active APC as well as the constant presence of Ni ions in the medium might, but need not, be reactive to cryptic peptides expressed upon Ni-induced aberrant protein processing.[59]

Individual T cell responses to Ni are usually highly polyclonal. However, in peripheral blood of patients with exceptionally strong hyperreactivity, a preference for BV17 expressing,

Ni-specific, $CD4^+$ T cells was observed, all restricted to the same HLA-DR allele.[71] In recent studies confirming these findings in larger numbers of patients, the $BV17^+$ cells were also identified in skin lesions.[72] Sequencing as well as TCR transfection and mutational studies of such $BV17^+$ receptors showed a further restriction of the repertoire regarding defined amino acid motifs (Arg-Asp) in the TCR β-chain CDR3 loops.[73] Functional replacement of Arg by His, but not by other amino acids pointed to this amino acid as a potential point of contact for Ni in the TCR. TCR α-chains pairing with such β-chains revealed a preference for AV1 elements, and several extremely similar, but not identical, TCR were identified in the same donor.[71,73]

Recent studies by Lu et al[74] indicate that in these cases Ni^{2+} ions, indeed, form part of the antigenic epitope by attaching to a histidine of the restricting HLA-DR β-chain as well as to amino acid side chains of as yet unidentified DR-associated peptides.

Concerning the biological function of these clones in vivo, one should keep in mind that none of them could be causally correlated to the pathology of Ni hypersensitivity reactions. Despite their identification in allergic skin leasions, these clones might functionally still belong to a negative regulatory population, particularly because none of them exhibited a typical Th1 cytokine profile.[43]

A New Model for TCR-Ni-MHC Contacts

A new model for TCR activation by Ni arose from the analysis of the Ni-specific, HLA-DR-restricted T cell clone SE9.[44] This clone was unusual in the sense that its DR-restriction was highly promiscuous with DR53 as the only definitely nonpresenting DR-allele. Moreover, in its AV22 and BV17 expressing TCR, the $BV17^+$ β-chain could be functionally replaced by almost any other TCR β-chain (even from murine T cells). For such TCR hybrids we occasionally observed partial, but never complete loss of Ni-reactivity.[44] Participation of MHC-bound peptides in defining Ni-specificity was excluded by showing that DR1 transfectants loaded with individually defined peptides via cotransfection with invariant chain mutants were indistinguishable in their potential to present Ni to clone SE9.[75] Instead, mutation of a conserved His residue in the DR1 β-chain, as well as of two hypervariable tyrosines in the CDR1 and CDR3 loops of the TCR α-chain, revealed three absolutely essential contact sites likely involved in Ni binding.[75] Hence, in this particular case, one Ni ion appears sufficient to form a coordination complex between TCR and MHC in a peptide-independent manner (Fig. 1). This complexation depends on the statistically rare situation that TCR and MHC molecules in the absence of Ni may already assume a conformation which supplies at least four potentially Ni-complexing amino acids in a perfect geometric arrangement. The binding of Ni, like a clamp, could add enough binding strength to the complex to allow for TCR phosphorylation, synapse formation and serial triggering.

Such a model is reminiscent of TCR-activation by superantigens which react with any TCR expressing certain BV-family members.[76] However, Ni ions in the above situation do not activate every AV22-expressing TCR, but require an idiotypic, N-nucleotide-determined amino acid in a unique TCR-alpha CDR3 sequence.[75] Although we have demonstrated this phenomenon for an admittedly unusual TCR, the principal TCR/MHC linkage via Ni-complexation may well be of general importance. The only requirement of the model is that due to positive thymic selection all peripheral T cells exhibit a basic natural "MHC-ness" allowing for short-term and instable TCR/MHC association. In a certain fraction of TCR/MHC combinations, these complexes may supply a perfect geometric arrangement of Ni coordination sites which, in the presence of Ni ions, stabilize the complex to a degree that allows for TCR signalling and T cell activation. In the majority of cases these coordination sites will probably locate to both of the TCR chains as well as to polymorphic MHC residues and will, therefore, not easily be identified.

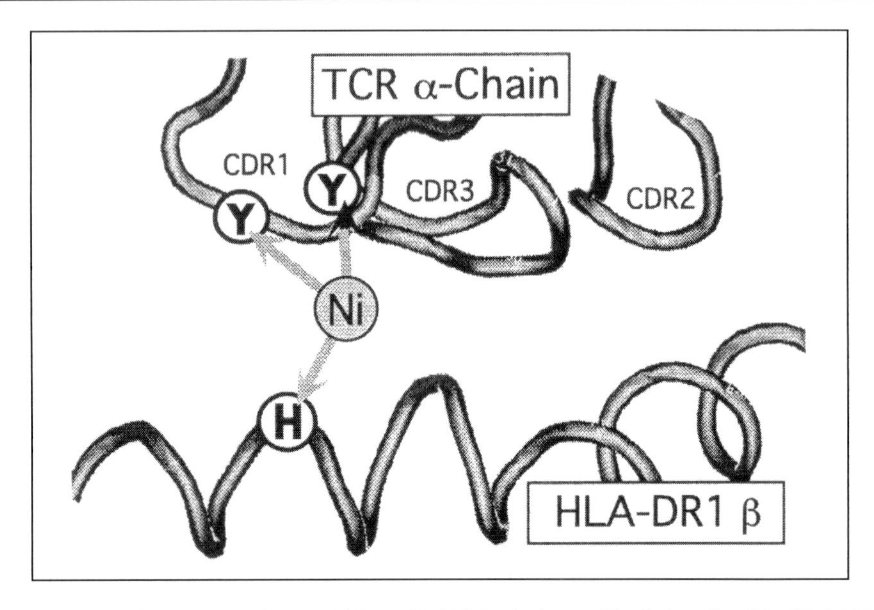

Figure 1. Potential contact sites for Ni in TCR and MHC for the human T cell clone SE9. The HLA-DR1 β-chain and TCR α-chain hypervariable loops of the crytallized complex of a hemagglutine-specific human TCR with HLA-DR1[7] are depicted. Highlighted are positions of His81 in the DR β-chain as well as Tyr30 in CDR1 and Tyr94 in CDR3 of the TCR α-chain. This view gives the best possible approximation of the potential Ni-coordination with HLA-DR1 and the TCR α-chain of clone SE9.[75]

Ni-Reactive $CD8^+$ T Cells

Whenever T cells from peripheral blood or from skin lesions of Ni-allergic patients are stimulated with Ni salts, the proliferating population consists of $CD4^+$ as well as of $CD8^+$ T cells.[43] In fact, Cavani et al[77] reported that CD8 proliferation correlates much better with a status of Ni-sensitization than CD4 proliferation, and that CD8 and CD4 cells could lyse keratinocytes in a Ni-specific manner.[78] However, whereas polyclonal Ni activation of peripheral blood cells always resulted in some 20% of proliferating CD8 T cells, cloning and particularly long-term in vitro culturing of these cells proved exceedingly difficult. Thus, comparatively little information is available on molecular details of Ni-recognition by class I MHC-restricted human T cells. However, some data on Ni-reactive CD8 clones and their cytolytic reactivity towards keratinocytes have been reported by the group of Cavani[78] and several long-term CTL clones were described by Moulon et al.[43] These clones were shown to express the skin homing receptor CLA, and upon stimulation to produce interferon (IFN)-γ and often also interleukin (IL)-4.[43] Most of the clones appeared to be clearly MHC-restricted, although the restricting HLA alleles have not been defined, and produced Ni-mediated cytolysis in classical chromium release assays. However, some clones were found which responded by proliferation as well as by cytolysis to Ni in a nonHLA-restricted manner.[43,79] In these cases, Ni-specific lysis was observed on a variety of HLA mis-matched target cells, including cells which either lack MHC class I or class II expression. The lysis is perforin mediated, and killing as well as proliferation require the permanent presence of Ni ions in the medium as well as metabolically active stimulator cells.[79] Transfection of the TCR of one of these clones into a murine hybridoma transferred HLA-unrestricted Ni-reactivity, indicating that the αβTCR is the only structure defining this unusual specificity. The antigenic epitopes recognized by these clones are not yet known, but since target cells can neither be fixed nor pulsed with Ni salts it is not even sure that Ni ions form part of the determinant. On the other hand, since these CTL

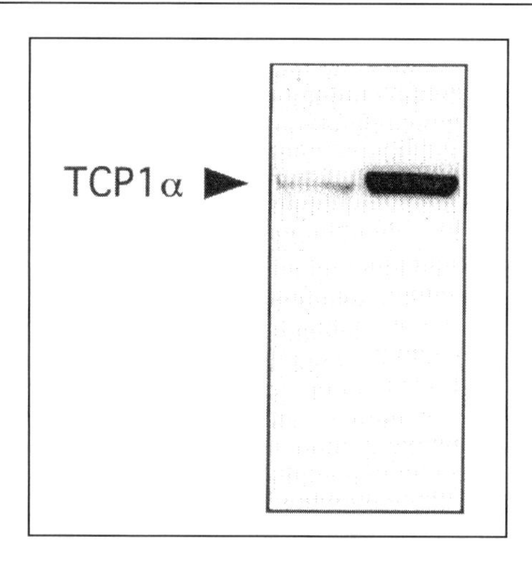

Figure 2. Enrichment of TCP1 α subunit by absorption to Ni-NTA agarose. Equivalent amounts of protein from a total Triton X-100 lysate of the B cell line WT47 (left) or the fraction of Ni-NTA agarose-enriched proteins thereof (right) were separated by SDS-PAGE and developed by Western blotting with a peroxidase-labelled antibody against the TCP1 α subunit (Stressgen, Victoria, Canada). Note the substantial enrichment of TCP1 by the Ni-beads.

react by interleukin secretion or cytolysis to Ni in the context of very many different cell types, they may constitute extremely effective enhancer systems in the induction or prevention of allergic responses.

Ni-Binding Proteins

Two questions in the context of Ni allergy cannot be answered via the identification of Ni-induced epitopes. These are (a) how do Ni ions get to the coordination sites in the TCR/MHC complexes, and (b) what determines that particularly nickel is such a potent allergen, in other words what supplies the "danger signal"[80] in the induction phase of Ni ACD. We argued that the low Ni concentrations in vivo will probably always exsist in the form of protein complexes. One protein which is long known for its Ni-binding properties is human serum albumin (HSA).[81] In fact, we could show that HSA-Ni complexes stimulate Ni-specific T cell clones in an HLA-restricted fashion, and also stimulate T cells in the peripheral blood of Ni-allergic individuals.[43,82] This stimulation does not depend on the species from which the serum albumin originates, except for dog albumin which lacks the property to bind Ni. From these data and experiments with fixed APC it became very unlikely that albumin-derived peptides themselves represented the MHC-associated binding sites for Ni. Since HSA is a prominent constituent in human skin and is able to cross the basal membrane to the dermis, it appeared as a highly probable candidate to carry Ni to the antigen presenting Langerhans cells in the skin. This prompted us to search for other Ni-binding proteins in and on cells.[83] Following enrichment of Ni-binding proteins from T- and B-cell lysates via Ni-NTA-agarose we identified several chaperones known as heat shock proteins (H.-J. Thierse, unpublished results). Figure 2 shows as one example the enrichment of the chaperonin containing TCP1(CCT) α subunit[84] as detected by Western blotting. These studies, besides identifying proteins other than HSA as possible Ni carriers, give a first hint in the direction of Ni-mediated danger signals. They demonstrate that allergenic metals, and probably other haptens, require more than the production of antigenic epitopes to induce an allergic response. Experiments to assay effects of Ni on expression and function of several heatshock proteins are under way.

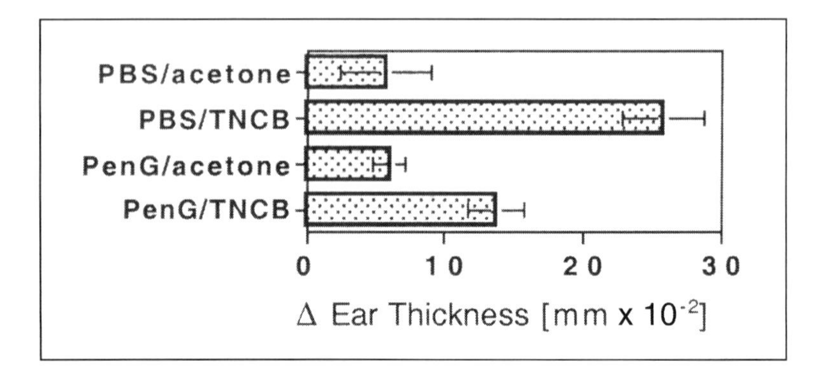

Figure 3. Reduced contact sensitivity in mice under treatment with penicillin G. C57BL/6 mice were skin-painted with 7% TNCB in acetone on the abdominal skin. On day 6, TNCB in acetone or acetone alone was applied on both ears, and ear thickness determined 24h later. From day -2 to day 6 mice were injected i.p. daily either with penicillin G (600 mg/kg) in phosphate buffered saline (PBS) or with PBS alone. Data represent means for three mice in each group with standard deviations indicated.

In the context of danger signals it is also important to note that it has recently become feasible to study Ni-allergic reactions in mice. After many years of unsuccessful attempts to induce contact sensitivity to $NiCl_2$ in mice, injection of Ni ions of higher oxidation status, i.e., Ni^{4+}, was found to sensitize mice to produce delayed-type hypersensitivity (DTH)-like reactions upon Ni^{2+} challenge.[85] This data demonstrates that the creation of Ni-determinants on APC in itself is not sufficient to induce a Ni-specific allergic reaction, but that it requires additional interactions of the metal ions with cellular metabolism. Such effects may be mediated through oxidative processes via higher oxidation levels of the metal ions, by interference with essential cellular constituents such as heatshock proteins or by many other processes. In any case, such "side effects" of haptens might also constitute a general property of nonmetal allergens.

Inhibition of Interferon Gamma by Penicillins

One astonishing example for metabolic "side-effects" of hapten allergens is the specific inhibition of IFN-γ secretion by penicillins. The phenomenon has been detected by Padovan et al[86] and has recently been investigated in detail by Dötze et al[87] (and unpublished results). These studies reveal that certain penicillins, i.e., penicillin G and V, besides producing HLA-associated penicilloyl peptides as allergenic epitopes,[45] are potent inhibitors of the secretion of IFN-γ by human and murine $CD4^+$ and $CD8^+$ T cells as well as by murine macrophages. The inhibition is specific for IFN-γ because the same penicillins do not affect the production of IL-2, IL-4, IL-5 or IL-10 (other interleukins not tested). Recent data by Brooks et al[88] claiming that penicillins destroy the epitopes for antibodies employed in the ELISA detection assays rather than blocking IFN-γ secretion were not reproduced in our hands. While the mechanisms underlying the inhibition of IFN-γ secretion are still a matter of research, the potential relevance of the effect for the developement of IgE-mediated penicillin allergy is apparent. A general reduction of IFN-γ during the induction of an immune response to penicillin would be expected to favour Th2 and IgE production. On the other hand, ACD as a typically IFN-γ dependent allergic reaction should be reduced under penicillin treatment. In fact, this was experimentally confirmed in a murine TNCB-induced ACD model[84] where ear swelling reactions in penicillin G treated mice was reduced by about 50% (see Fig. 3).

In contrast, penicillin G should act counterproductively in the control of viral infections where IFN-γ is required for virus elimination. Indeed, preliminary data in cooperation with C.

Leipner at the University of Jena/Germany with coxsackie type B virus-induced myocarditis appear to indicate reduced survival in penicillin G treated versus control animals. Independent of their theoretical impact, these findings put a serious caveat on the penicillin G treatment of patients suffering from undefined infections. Future studies will concentrate on identifying antibiotics which lack the IFN-γ blocking properties in vivo, as well as on the elucidation of the molecular mechanisms underlying the inhibition of IFN-γ secretion by penicillins.

Noncovalent MHC Association of Drugs

Ever since the times of Karl Landsteiner, haptens have been defined as low molecular chemicals which need to be covalently linked to proteins before inducing an immune response. Nonreactive chemicals and drugs exhibiting allergenic properties were, therefore, termed pro-haptens. It was assumed, and in several cases experimentally proven, that pro-haptens needed to be transformed into reactive metabolites prior to inducing immune responses or allergic reactions. However, recent findings by the group of Pichler support a model according to which certain drugs may also be presented to T cells in a noncovalent MHC-drug association. The authors provide evidence that, for example, sulfamethoxazole (SMX) may not only be metabolized to nitroso compounds which react covalently with proteins including MHC, but also may associate noncovalently with MHC/peptide complexes.[89,90] In the latter situation the permanent presence of SMX in the medium is required, but glutaraldehyde-fixed APC can present the drug. In that sense SMX acts similarly to metal ions in a TCR/MHC complex stabilized by noncovalent bonds to the hapten. Support for this model comes from data showing that T cells reactive to metabolite-induced epitopes do not cross-react with noncovalently bound SMX.[91] Noncovalent drug-MHC interactions, thus, may constitute a novel type of MHC-restricted drug epitope.

TCR Transfectants As Tools to Study Hapten Recognition

Many studies concerning TCR activation by haptens and in particular TCR mutation experiments are conducted in T cell fusion-hybridomas or in TCR transfectants, respectively.[32,44,92,93] In general, such cells are true images of the genuine T cell clones in terms of antigen specificity. For this reason it is generally assumed that TCR assembly, specificity and signaling pathways in hybridomas resemble those in clonal T cells in vitro as well as in vivo. Recent results from our laboratory shed some doubt on this assumption (S.Hellwig, unpublished data). By in vitro stimulation with an H-2K^b binding TNP-peptide we have induced a murine CTL clone (E6)[31] and fused it with the TCR-deficient thymoma BW5147. The resulting T cell hybridoma (HyE6) exhibited no reactivity for the inducing antigen (M4L-TNP), but retained specificity for a crossreactive TNP-peptide (O4-TNP).

Clone E6 expresses one TCR β-chain (BV16) and two α-chains (AV3 and AV17) on its surface, but only AV17 is transported to the surface of the hybridoma HyE6 (see Fig. 4). Hence, the loss of specificity seems to be related to the loss of one of the two TCR. However, mRNA and even intracellular protein (not shown) of the AV3$^+$ α-chain are present in HyE6. Individual transfections of the two α-chain genes in combination with the BV16$^+$ β-chain gene revealed a total lack of surface expression of the AV3/BV16 combination while the AV17/BV16 heterodimer was expressed well (Fig. 4). and exhibited identical specificity as HyE6. Successful pairings were observed for the AV3 chain with several other murine TCR β-chains (not shown) and TCR expression was also induced upon supertransfection of the AV3/BV16 transfectant with an unrelated BV2$^+$ β-chain (Fig. 4). This indicates a particular property of the AV3$^+$ α-chain of clone E6 not to pair in hybridomas with the β-chain of the same clone, even in the presence of a second (AV17) α-chain. However, the fact that clone E6 does express the AV3$^+$ α-chain in a complex with the AV17/BV16 TCR strongly suggests a different organization of TCR molecules on T cell clones and hybridomas.

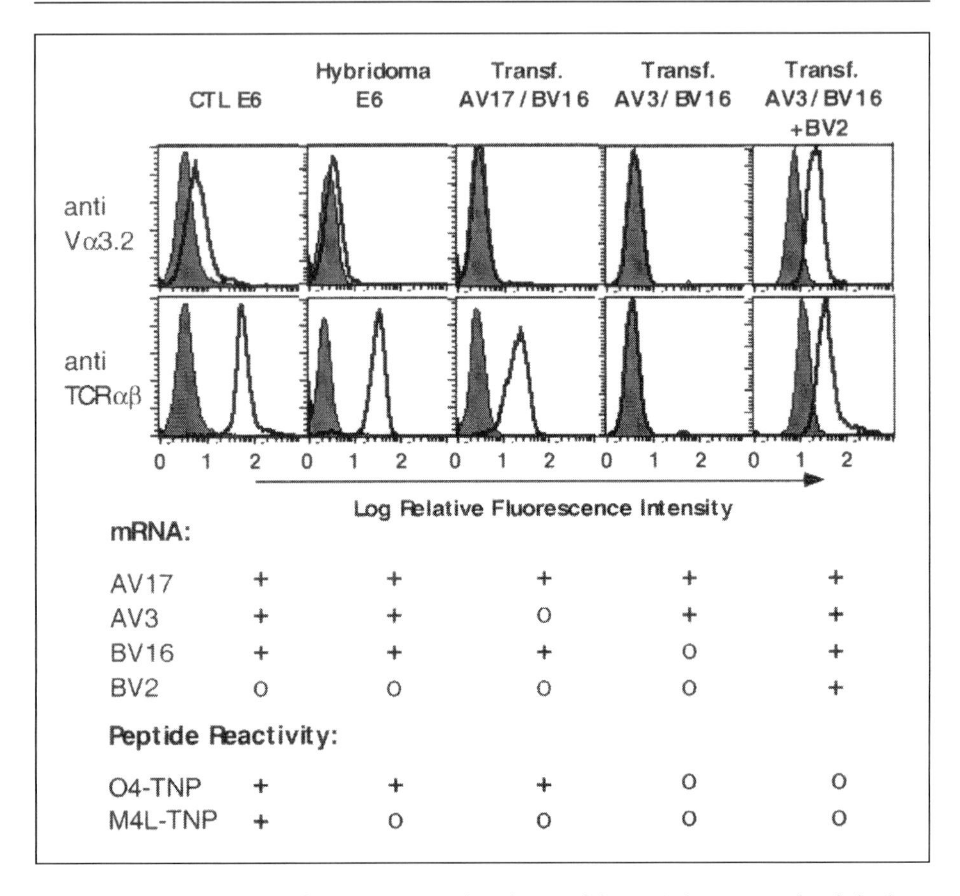

Figure 4. Cell surface stainings for TCR Vα3.2 and total TCR of clone E6, the corresponding hybridoma HyE6, and different TCR transfectants. Cells were stained with antibody RR3-16 for Vα3.2 or H57-597 for TCRαβ and analyzed in a fluorescence activated cell sorter. Open histograms show stanings with FITC-labeled specific antibodies, closed histograms are isotype controls. Indicated below are the presence of correctly spliced mRNA for the various TCR chains, as determined by PCR and DNA sequencing, and the reactivities to peptides M4L-TNP or O4-TNP. Peptide sequences are SMQK*FGEL (M4L) and SIIK*FEKL (O4)[31] with asterisks indicating TNP modification.

Conclusions

Haptens as T cell antigens deserve attention under three important aspects. Firstly, in many cases haptens are highly potent inducers of allergic disorders, particularly of type IV reactions such as ACD. Understanding the mechanisms underlying their allergenic potential eventually should help to establish in vitro tests to differentiate potentially strong from poor allergens. Secondly, in many cases hapten-specific T cells have been shown to interact with hapten-modified peptides in the presenting MHC binding groove absolutely identical to T cells specific for nominal peptide antigens. In such cases the hapten system has the great advantage over most MHC-peptide combinations that hapten/MHC complexes can be visualized and quantified on living cells with hapten-specific antibodies independent of peptide sequence or MHC haplotype. Finally, as mentioned above, there are several conditions where the molecular mechanisms of hapten interaction with MHC and TCR are in a very early state of understanding. The finding of noncovalent drug association with MHC as well as many aspects of T cell interactions with metal ions indicate that some of the many routes to T cell activation may still be waiting to be uncovered. The most important lesson is, however, that haptens are not just part of chemists' games in immunology, but a very real threat to the immune system in so-called natural conditions as well as particularly in the modern technological environment.

Acknowledgements

The authors' research was funded by the Deutsche Forschungsgemeinschaft (SFB388 and We 379/9-1) and the German Federal Ministry of Education and Research (BMBF) in the clinical research group "pathomechanisms of allergic inflammation" (FKZ 01GC0102).

References

1. Eisen HN. Specificity and degeneracy in antigen recognition: Yin and yang in the immune system. Annu Rev Immunol 2001; 19:1-21.
2. Zinkernagel RM, Doherty PC. Restriction of in vitro T cell-mediated cytotoxicity in lymphocytic choriomeningitis within a syngeneic or semiallogeneic system. Nature 1974; 248:701-702.
3. Shearer GM. Cell-mediated cytotoxicity to trinitrophenyl-modified syngeneic lymphocytes. Eur J Immunol 1974; 4:527-533.
4. Haskins K, Kappler J, Marrack P. The major histocompatibility complex-restricted antigen receptor on T cells. Annu Rev Immunol 1984; 2:51-66.
5. Buus S, Sette A, Colon SM et al. The relation between major histocompatibility complex (MHC) restriction and the capacity of IA to bind immunogenic peptides. Science 1987; 235:1353-1358.
6. Garcia KC, Tallquist MD, Pease LR et al. Alpha-beta-T-cell receptor interactions with syngeneic and allogeneic ligands—affinity measurements and crystallization. Proc Natl Acad Sci USA 1997; 94:13838-13843.
7. Hennecke J, Carfi A, Wiley DC. Structure of a covalently stabilized complex of a human alphabeta T-cell receptor, influenza HA peptide and MHC class II molecule, HLA-DR1. EMBO J 2000; 19:5611-5624.
8. Margulies DH. Interactions of TCRs with MHC-peptide complexes - a quantitative basis for mechanistic models. Curr Opin Immunol 1997; 9:390-395.
9. Germain RN, Stefanova I. The dynamics of T cell receptor signaling: Complex orchestration and the key roles of tempo and cooperation. Annu Rev Immunol 1999; 17:467-522.
10. Lebrec H, Kerdine S, Gaspard I et al. Th(1)/Th(2) responses to drugs. Toxicology 2001; 158:25-29.
11. Kimber I, Dearman RJ. Allergic contact dermatitis: The cellular effectors. Contact Dermatitis 2002; 46:1-5.
12. Girolomoni G, Sebastiani S, Albanesi C et al. T-cell subpopulations in the development of atopic and contact allergy. Curr Opin Immunol 2001; 13:733-737.
13. Pohlit H, Haas W, von Boehmer H. Cytotoxic T cell responses to haptenated cells. I. Primary, secondary and long-term cultures. Eur J Immunol 1979; 9:681-690.
14. Levy RB, Shearer GM. Cell-mediated lympholysis responses against autologous cells modified with haptenic sulfhydryl reagents. IV. Self-determinants recognized by wild-type anti-H-2Kb and H-2Db-restricted cytotoxic T cells specific for sulfhydryl and amino-reactive haptens are absent in certain H-2 mutant strains. J Immunol 1982; 129:1525-1529.
15. Schmitt-Verhulst A-M, Pettinelli CB, Henkart PA et al. H-2-restricted cytotoxic effectors generated in vitro by the addition of trinitrophenyl-conjugated soluble proteins. J Exp Med 1978; 174:352-368.
16. Ciavarra R, Forman J. Cells treated with trinitrobenzene sulfonic acid express an antigenic determinant recognized by cytotoxic effector cells that is not detected on cells coated with trinitrophenylated proteins. J Immunol 1980; 124:713-718.
17. Ortmann B, Martin S, von Bonin A et al. Synthetic peptides anchor T cell-specific TNP epitopes to MHC antigens. J Immunol 1992; 148:1445-1450.
18. Nalefski EA, Rao A. Nature of the ligand recognized by a hapten- and carrier-specific, MHC-restricted T cell receptor. J Immunol 1993; 150:3806-3816.
19. Weltzien HU, Moulon C, Martin S et al. T cell immune responses to haptens. Structural models for allergic and autoimmune reactions. Toxicology 1996; 107:141-151.
20. von Bonin A, Ortmann B, Martin S et al. Peptide-conjugated hapten groups are the major antigenic determinants for trinitrophenyl-specific cytotoxic T cells. Int Immunol 1992; 4:869-874.

21. Martin S, von Bonin A, Fessler C et al. Structural complexity of antigenic determinants for class I MHC- restricted, hapten-specific T cells. Two qualitatively differing types of H-2Kb-restricted TNP epitopes. J Immunol 1993; 151:678-687.
22. Kohler J, Hartmann U, Grimm R et al. Carrier-independent hapten recognition and promiscuous MHC restriction by CD4 T cells induced by trinitrophenylated peptides. J Immunol 1997; 158:591-597.
23. Martin S, Weltzien HU. T cell recognition of haptens, a molecular view. Int Arch Allergy Immunol 1994; 104:10-16.
24. Rammensee H-G. MHC ligands and peptide motifs: First listing. Immunogenetics 1995; 41:178-228.
25. Kohler J, Martin S, Pflugfelder U et al. Cross-reactive trinitrophenylated peptides as antigens for class II major histocompatibility complex-restricted T cells and inducers of contact sensitivity in mice. Limited T cell receptor repertoire. European Journal of Immunology 1995; 25:92-101.
26. Martin S, Lappin MB, Kohler J et al. Peptide immunization indicates that CD8+ T cells are the dominant effector cells in trinitrophenyl-specific contact hypersensitivity. J Invest Dermatol 2000; 115:260-266.
27. Simon JC, Martin S. Rapid Induction of Hapten-Specific Tc1 Cells in vivo by i.c. Injections of Trinitrophenol-Modified Dendritic Cells. Int Arch Allergy Immunol 2001; 124:221-222.
28. Martin S, DeLattre V, Leicht C et al. A high frequency of allergen-specific CD8+ Tc1 cells is associated with the murine immune response to the contact sensitizer trinitrophenyl. Exp Dermatol 2003; 12:78-85.
29. Weltzien HU, Padovan E. Molecular features of penicillin allergy. J Invest Dermatol 1998; 110:203-206.
30. Martin S, Ortmann B, Pflugfelder U et al. Role of hapten-anchoring peptides in defining hapten-epitopes for MHC-restricted cytotoxic T cells. Cross-reactive TNP-determinants on different peptides. J Immunol 1992; 149:2569-2575.
31. Preckel T, Breloer M, Kohler H et al. Partial agonism and independent modulation of T cell receptor and CD8 in hapten-specific cytotoxic T cells. Eur J Immunol 1998; 28:3706-3718.
32. von Bonin A, Martin S, Plaga S et al. Purified MHC class I molecules present hapten-conjugated peptides to TNP/H-2Kb-specific T cell hybridomas. Immunol Lett 1993; 35:63-68.
33. Martin S, Niedermann G, Leipner C et al. Intracellular processing of hapten-modified protein for MHC class I presentation: Cytoplasmic delivery by pH-sensitive liposomes. Immunol Lett 1993; 37:97-102.
34. Uetrecht J. Metabolism of drugs by activated leukocytes: Implications for drug- induced lupus and other drug hypersensitivity reactions. Adv Exp Med Biol 1991; 283:121-132.
35. Baron JM, Merk HF. Drug metabolism in the skin. Curr Opin Allergy Clin Immunol 2001; 1:287-291.
36. Kalish RS, Wood JA, LaPorte A. Processing of urushiol (poison ivy) hapten by both endogenous and exogenous pathways for presentation to T cells in vitro. J Clin Invest 1994; 93:2039-2047.
37. Naisbitt DJ, Gordon SF, Pirmohamed M et al. Antigenicity and immunogenicity of sulphamethoxazole: Demonstration of metabolism-dependent haptenation and T-cell proliferation in vivo. Br J Pharmacol 2001; 133:295-305.
38. Hansson C, Ahlfors S, Bergendorff O. Concomitant contact-dermatitis due to textile dyes and to color film developers can be explained by the formation of the same hapten. Contact Dermatitis 1997; 37:27-31.
39. Wulferink M, Gonzalez J, Goebel C et al. T cells ignore aniline, a prohapten, but respond to its reactive metabolites generated by phagocytes: Possible implications for the pathogenesis of toxic oil syndrome. Chem Res Toxicol 2001; 14:389-397.
40. Iglesias A, Hansen-Hagge T, Von Bonin A et al. Increased frequency of 2,4,6-trinitrophenyl (TNP)-specific, H-2b-restricted cytotoxic T lymphocyte precursors in transgenic mice expressing a T cell receptor beta chain gene from an H-2b-restricted, TNP-specific cytolytic T cell clone. Eur J Immunol 1992; 22:335-341.
41. Hochgeschwender U, Simon HG, Weltzien HU et al. Dominance of one T-cell receptor in the H-2Kb/TNP response. Nature 1987; 326:307-309.
42. Kempkes B, Palmer E, Martin S et al. Predominant T cell receptor gene elements in TNP-specific cytotoxic T cells. J Immunol 1991; 147:2467-2473.
43. Moulon C, Wild D, Dormoy A et al. MHC-dependent and -independent activation of human nickel-specific CD8+ cytotoxic T cells from allergic donors. J Invest Dermatol 1998; 111:360-366.

44. Vollmer J, Weltzien HU, Gamerdinger K et al. Antigen contacts by Ni-reactive TCR: Typical αβ chain cooperation versus α chain-dominated specificity. Int Immunol 2000; 12:1723-1731.
45. Padovan E, Bauer T, Tongio MM et al. Penicilloyl peptides are recognized as T cell antigenic determinants in penicillin allergy. Eur J Immunol 1997; 27:1303-1307.
46. Garcia KC, Degano M, Stanfield RL et al. An alpha-beta-T-cell receptor structure at 2.5 angstrom and its orientation in the TCR-MHC complex. Science 1996; 274:209-219.
47. Kessler B, Michielin O, Blanchard CL et al. T cell recognition of hapten - Anatomy of T cell receptor binding of a H-2Kd-associated photoreactive peptide derivative. J Biol Chem 1999; 274:3622-3631.
48. Doucey MA, Legler DF, Boucheron N et al. CTL activation is induced by cross-linking of TCR/MHC-peptide-CD8/p56lck adducts in rafts. Eur J Immunol 2001; 31:1561-1570.
49. Preckel T, Grimm R, Martin S et al. Altered hapten ligands antagonize trinitrophenyl-specific cytotoxic T cells and block internalization of hapten-specific receptors. J Exp Med 1997; 185:1803-1813.
50. Preckel T, Hellwig S, Pflugfelder U et al. Clonal anergy induced in a CD8+ hapten-specific cytotoxic T-cell clone by an altered hapten-peptide ligand. Immunology 2001; 102:8-14.
51. Kreiss K, Miller F, Newman LS et al. Chronic beryllium disease - from the workplace to cellular immunology, molecular immunogenetics, and back. Clin Immunol Immunopathol 1994; 71:123-129.
52. Lison D, Lauwerys R, Demedts M et al. Experimental research into the pathogenesis of cobalt/hard metal lung disease. Eur Respir J 1996; 9:1024-1028.
53. Schuppe HC, Ronnau AC, von Schmiedeberg S et al. Immunomodulation by heavy metal compounds. Clin Dermatol 1998; 16:149-157.
54. Sinigaglia F. The molecular basis of metal recognition by T cells [published erratum appears in J Invest Dermatol 1994; 103:41]. J Invest Dermatol 1994; 102:398-401.
55. Budinger L, Hertl M. Immunologic mechanisms in hypersensitivity reactions to metal ions: An overview. Allergy 2000; 55:108-115.
56. Fausto da Silva JJR, Williams RJP. The biological chemistry of the elements. The inorganic chemistry of life. 2nd ed. Oxford, UK: Oxford University Press, 2001:39-51.
57. Romagnoli P, Labhardt AM, Sinigaglia F. Selective interaction of Ni with an MHC-bound peptide. EMBO J 1991; 10:1303-1306.
58. Richeldi L, Sorrentino R, Saltini C. HLA-DPB1 glutamate 69: A genetic marker of beryllium disease. Science 1993; 262:242 - 244.
59. Griem P, Vonvultee C, Panthel K et al. T-cell cross-reactivity to heavy-metals - identical cryptic peptides may be presented from protein exposed to different metals. Eur J Immunol 1998; 28:1941-1947.
60. Wang Z, White PS, Petrovic M et al. Differential susceptibilities to chronic beryllium disease contributed by different Glu69 HLA-DPB1 and -DPA1 alleles. J Immunol 1999; 163:1647-1653.
61. Potolicchio I, Festucci A, Hausler P et al. HLA-DP molecules bind cobalt: A possible explanation for the genetic association with hard metal disease. E J Immunol 1999; 29:2140-2147.
62. Griem P, Panthel K, Kalbacher H et al. Alteration of a model antigen by Au(III) leads to T cell sensitization to cryptic peptides. E J Immunol 1996; 26:279-287.
63. Griem P, Wulferink M, Sachs B et al. Allergic and autoimmune reactions to xenobiotics: How do they arise? Immunol Today 1998; 19:133-141.
64. Sinigaglia F, Scheidegger D, Garotta G et al. Isolation and characterization of Ni-specific T cell clones from patients with Ni-contact dermatitis. J Immunol 1985; 135:3929-3932.
65. Kapsenberg ML, Res P, Bos JD et al. Nickel-specific T lymphocyte clones derived from allergic nickel-contact dermatitis lesions in man: Heterogeneity based on requirement of dendritic antigen-presenting cell subsets. European J Immunol 1987; 17:861-865.
66. Kapsenberg ML, Bos JD, Wierenga EA. T cells in allergic responses to haptens and proteins. Springer Semin Immunopathol 1992; 13:303-314.
67. Moulon C, Vollmer J, Weltzien HU. Characterization of processing requirements and metal cross-reactivities in T cell clones from patients with allergic contact dermatitis to nickel. Eur J Immunol 1995; 25:3308-3315.
68. Pistoor FHM, Kapsenberg ML, Bos JD et al. Cross-reactivity of human nickel-reactive T lymphocyte clones with copper and palladium. J Invest Dermatol 1995; 105:92-95.
69. Werfel T, Hentschel M, Renz H et al. Analysis of the phenotype and cytokine pattern of blood- and skin-derived nickel specific T cells in allergic contact dermatitis. Int Arch Allergy Immunol 1997; 113:384-386.

70. Emtestam L, Zetterquist H, Olerup O. HLA-DR, -DQ and -DP alleles in nickel, chromium, and/or cobalt-sensitive individuals: Genomic analysis based on restriction fragment length polymorphism. J Invest Dermatol 1993; 100:271-274.
71. Vollmer J, Fritz M, Dormoy A et al. Dominance of the BV17 element in nickel-specific human T cell receptors relates to severity of contact sensitivity. European J Immunol 1997; 27:1865-1874.
72. Budinger L, Neuser N, Totzke U et al. Preferential usage of TCR-Vbeta17 by peripheral and cutaneous T cells in nickel-induced contact dermatitis. J Immunol 2001; 167:6038-6044.
73. Vollmer J, Weltzien HU, Moulon C. TCR reactivity in human nickel allergy indicates contacts with complementarity-determining region 3 but excludes superantigen-like recognition. J Immunol 1999; 163:2723-2731.
74. Lu L, Vollmer J, Moulon C et al. Components of the ligand for a Ni++ reactive human T cell clone. J Exp Med 2003; 197:567-574.
75. Gamerdinger K, Moulon C, Karp DR et al. A new type of metal recognition by human T cells: contact residues for peptide-independent bridging of TCR and MHC by nickel. J Exp Med 2003; 197:1345-1353.
76. Li HM, Llera A, Malchiodi EL et al. The structural basis of T cell activation by superantigens. Annu Rev Immunol 1999; 17:435-466.
77. Cavani A, Mei D, Guerra E et al. Patients with allergic contact-dermatitis to nickel and nonallergic individuals display different nickel-specific T-cell responses—evidence for the presence of effector CD8+ and regulatory CD4+ T-cells. J Invest Dermatol 1998; 111:621-628.
78. Traidl C, Sebastiani S, Albanesi C et al. Disparate cytotoxic activity of nickel-specific CD8+ and CD4+ T cell subsets against keratinocytes. J Immunol 2000; 165:3058-3064.
79. Moulon C, Choleva Y, Thierse H-J et al. TCR transfection shows non-HLA-restricted recognition of nickel by CD8+ human T cells to be mediated by $\alpha\beta$ T cell receptors. J Invest Dermatol 2003; 121:496-501.
80. Matzinger P. The danger model: A renewed sense of self. Science 2002; 296:301-305.
81. Bal W, Christodoulou J, Sadler PJ et al. Multi-metal binding site of serum albumin. J Inorg Biochem 1998; 70:33-39.
82. Thierse H-J, Moulon C, Allespach Y et al. Metal-protein complex mediated transport and delivery of Ni2+ to TCR/MHC contact sites in nickel-specific human T cell activation. J Immunol 2004; 172:1926-1934.
83. Thierse HJ, Moulon C, Wild D et al. Analysis of nickel-binding proteins in human T cells. In: Mackiewicz A, Kurpisz M, Zeromski J, eds. EFIS 2000, Proceedings 14th European Immunology Meeting Poznan, Poland. Bologna, Italy: Monduzzi Editore, 2000:727-732.
84. Leroux MR, Hartl FU. Protein folding: Versatility of the cytosolic chaperonin TRiC/CCT. Curr Biol 2000; 10:R260-264.
85. Artik S, von Vultee C, Gleichmann E et al. Nickel allergy in mice: Enhanced sensitization capacity of nickel at higher oxidation states. J Immunol 1999; 163:1143-1152.
86. Padovan E, von Greyerz S, Pichler WJ et al. Antigen-dependent and -independent IFN-gamma modulation by penicillins. J Immunol 1999; 162:1171-1177.
87. Dötze A, Martin S, Modolell M et al. Penicillin: An anti-inflammatory drug? (abstract). Immunobiology 2000; 203:568-568.
88. Brooks BM, Flanagan BF, Thomas AL et al. Penicillin conjugates to interferon-gamma and reduces its activity: A novel drug-cytokine interaction. Biochem Biophys Res Comm 2001; 288:1175-1181.
89. Schnyder B, Burkhart C, Schnyder-Frutig K et al. Recognition of sulfamethoxazole and its reactive metabolites by drug-specific CD4+ T cells from allergic individuals. J Immunol 2000; 164:6647-6654.
90. Britschgi M, von Greyerz S, Burkhart C et al. Molecular aspects of drug recognition by specific T cells. Current Drug Targets 2003; 4:1-11.
91. Burkhart C, von Greyerz S, Depta JP et al. Influence of reduced glutathione on the proliferative response of sulfamethoxazole-specific and sulfamethoxazole-metabolite-specific human CD4+ T-cells. Br J Pharmacol 2001; 132:623-630.
92. Ulivieri C, Peter A, Orsini E et al. Defective signaling to Fyn by a T cell antigen receptor lacking the alpha-chain connecting peptide motif. J Biol Chem 2001; 276:3574-3580.
93. Stotz SH, Bolliger L, Carbone FR et al. T cell receptor (TCR) antagonism without a negative signal: Evidence from T cell hybridomas expressing two independent TCRs. J Exp Med 1999; 189:253-263.

CHAPTER 3

Langerhans Cell Migration and the Induction Phase of Skin Sensitization

Marie Cumberbatch, Rebecca J. Dearman, Christopher E.M. Griffiths, Richard W. Groves and Ian Kimber

Summary

Langerhans cells (LC) are members of a wide family of bone marrow derived, immunoactive dendritic cells (DC). LC reside in the epidermis where they are regarded as sentinels of the adaptive immune system with responsibility for surveying changes in the microenvironment and forming a trap for external antigen. In response to local antigen (including contact allergen) and/or skin trauma, LC are mobilized, induced to leave the skin and migrate via afferent lymphatics to draining lymph nodes. During this journey LC are subject to a functional maturation such that they differentiate from what are essentially antigen recognition and processing cells in the skin, to mature immunocompetent DC that are able to present antigen effectively to responsive T lymphocytes. The migration of LC from the skin is a very complex process. Coordinated interactions are required between cytokines and chemokines that together orchestrate the changes in LC necessary for successful trafficking from the skin to regional lymph nodes. If LC migration is compromised then skin sensitization fails to develop, or is sub-optimal. Here we examine in detail the molecular and cellular events that initiate and regulate LC migration during the induction phase of skin sensitization.

Introduction

The acquisition of skin sensitization is characterized by, and dependent upon, the stimulation of a specific immune response in lymph nodes draining the site of exposure to contact allergen. This in turn requires that antigen is recognized and processed in the skin and delivered in an immunogenic form to responsive T lymphocytes in regional lymph nodes. These are functions performed by epidermal Langerhans cells (LC), which form part of the wider family of dendritic cells (DC) that collectively initiate and orchestrate adaptive immune responses.[1,2]

Within the epidermis LC exhibit a unique phenotype displaying a dendritic morphology with dendrites interdigitating between adjacent keratinocytes. They contain characteristic Birbeck granules and express constitutively a variety of membrane determinants including major histocompatibility complex (MHC) class I and class II molecules, E-cadherin, Fc receptors for both IgG and IgE (FcγRII; CD32 and FcεRII; CD23), various cytokine and chemokine receptors and a range of adhesion molecules.[3,4] They derive from myeloid precursors and/or from a common DC-committed progenitor population in the bone marrow,[5-7] and gain access to the epidermis via the blood.[8,9]

Immune Mechanisms in Allergic Contact Dermatitis, edited by Andrea Cavani and Giampiero Girolomoni. ©2005 Eurekah.com.

It is well established that in response to skin sensitization a proportion of LC local to the site of exposure (some of which bear detectable levels of antigen) become mobilized and migrate from the epidermis, via afferent lymphatics, to draining lymph nodes.[10-15] During migration these cells are subject to a functional maturation and by the time of their arrival in lymph nodes they have assumed the characteristics of mature DC. This differentiation, which is mediated in part by epidermal cytokines, serves to transform LC with the characteristics of antigen processing cells into immunostimulatory DC that have the ability to present antigen, either directly or indirectly, to responsive T lymphocytes.[16-19] The biology of LC, and the changes to which they are subject following activation, have been the subject of several reviews.[9,20-22] Here, however, attention will focus on the migration of these cells from the skin during the induction phase of contact sensitization.

Before engaging in that discussion it is necessary to make two points. First, although the theme of this article is LC, it is important to acknowledge that other forms of DC are found within the skin and that some of these cells, and also dermal macrophages, may have the ability to traffic to regional lymph nodes and in some circumstances influence the induction phase of skin sensitization.[23-25] Second, although LC are often regarded as being sentinels of the immune system, with responsibility for sampling changes in the antigenic environment within the skin, it is clear that encounter with antigen is not necessary for their mobilization. Thus, other forms of skin trauma, including UVB irradiation and chemical induced skin irritation are both associated with LC migration,[26-29] although the molecular mechanisms through which LC mobilization is achieved may vary according to the nature of the stimulus.[29] Our view currently is that LC mobilization will be stimulated under any conditions that are associated with an increased local availability of proinflammatory cytokines and that some degree of skin trauma (either from the antigenic stimulus itself, or from concomitant perturbation of the skin) may be required for optimal acquisition of sensitization.

Cytokines and the Initiation of LC Mobilization

Epidermal cytokines play an essential role in the initiation and regulation of LC migration from the skin. Attention focused initially on tumor necrosis factor (TNF)-α, an inducible product of keratinocytes that is known to be up-regulated following skin sensitization.[30] Studies in mice revealed that (intradermal) injection of recombinant homologous TNF-α was able to cause a rapid reduction in the frequency of LC in the epidermis local to the site of exposure, and that this was accompanied somewhat later by an accumulation of DC in draining lymph nodes.[31,32] More recently, it has been possible to demonstrate that homologous TNF-α causes an identical migration of human LC.[33-35] Not only is TNF-α able to initiate migration, it is also required for the mobilization of LC in response to skin sensitization. Systemic treatment of mice with neutralizing anti-TNF-α antibody prior to skin sensitization was shown to cause a substantial reduction in DC accumulation in draining lymph nodes (associated with an inhibition of contact hypersensitivity).[36] Indirect evidence suggests that TNF-α is also required for contact allergen-induced LC migration in humans. Studies were performed in volunteers with recombinant human lactoferrin (LF), a protein known to compromise the production of TNF-α in the skin.[37] Topical treatment of volunteers with diphenylcyclopropenone (DPC; a potent contact allergen) was found to cause a significant reduction in the frequency of LC local to the site of exposure. If, however, prior to sensitization, subjects were exposed at the same site to LF then the allergen-induced mobilization of LC was inhibited significantly, or prevented completely. Under the same experimental conditions LF failed to affect the integrity of LC migration provoked by exposure of volunteers to TNF-α.[34] In mice also, homologous LF causes an inhibition of allergen-induced, but not TNF-α-induced, migration,[38] as does dexamethasone, a transcriptional inhibitor of TNF-α.[39] It is clear that in both man and mouse TNF-α plays a pivotal role in the initiation of LC migration. Several lines of evidence indicate that the signal provided by this cytokine is delivered exclusively via the species-selective type 2 receptor for TNF-α (TNF-R2). Thus, heterologous (human) TNF-α fails in mice to provoke either LC migration or the accumulation of DC in draining lymph nodes,[31,32] and LC migration is inhibited markedly in TNF-R2 gene knockout mice, but not in those lacking TNF-R1.[40,41] Moreover, the consensus is that LC display only TNF-R2.[42-44]

A second cytokine that plays an essential role in allergen-induced LC migration is interleukin (IL)-1β. This cytokine is expressed constitutively by LC, and in mouse epidermis at least appears to be produced exclusively by these cells.[30] Like TNF-α, IL-1β when administered to mice by intradermal injection is able to induce both LC migration and the accumulation of DC in draining lymph nodes, albeit with somewhat slower kinetics.[45] Not unexpectedly, mobilization of LC by IL-1β is dependent upon intact function of IL-1RI, the signal transducing receptor for IL-1 that is known to be expressed by both LC and keratinocytes.[46,47] Interleukin 1β is also required for the stimulation by chemical allergens of LC migration.[48] In fact the normal mobilization of LC in response to skin sensitization requires both TNF-α and IL-1β; allergen-induced migration being compromised markedly if animals are pre-treated with neutralizing antibodies specific for either cytokine.[48] Initially it was thought that the action of IL-1β might be solely to stimulate the production by keratinocytes of TNF-α (a known property of IL-1β),[49] and certainly such an explanation would be consistent with the fact that the kinetics of LC mobilization in response to IL-1β were somewhat slower than those observed with TNF-α.[45] However, it was found that while, as expected, anti-TNF-α antibody was able to inhibit migration induced in mice by IL-1β, the converse was also the case. Thus, pre-treatment of mice with neutralizing anti-IL-1β antibody caused a significant inhibition of LC migration and DC accumulation induced by intradermal TNF-α. These data demonstrate that the mandatory role played by IL-1β during the initiation of LC activation extends beyond provision of a stimulus for local production of TNF-α. The conclusion drawn was that LC-derived IL-1β actually fulfils two independent requirements; one being to stimulate in paracrine fashion the synthesis and secretion of TNF-α by adjacent keratinocytes, the other being to provide an autocrine signal to LC themselves.[50] On this basis the interpretation is that when administered alone IL-1β is able to induce migration because, in addition to delivering a signal directly to LC, it will provoke the production of TNF-α. In the case of TNF-α it is believed that when this cytokine is administered alone there is available sufficient constitutive IL-1β to support migration by provision of a signal to LC.[50] The need for the de novo production of TNF-α for the stimulation of migration by IL-1β is illustrated by results of experiments performed in mice and in humans with LF. In both cases it was found that pre-treatment with LF was able to inhibit LC migration induced by local injection of IL-1β.[37,38,51] As these latter data imply, there is now evidence that IL-1β is able to cause LC migration in humans,[51] indicating that there is substantial conservation between species with respect to the molecular mechanisms that control LC function.

Further support for the important role of IL-1β in the regulation of LC function has derived from investigations of caspase-1, the IL-1β converting enzyme (ICE). It was found that in caspase-1 deficient mice neither contact allergen, nor TNF-α, were able to induce LC migration, although migration was normal in response to IL-1β.[52] Furthermore, studies conducted in vitro with skin organ cultures revealed that a peptide inhibitor of caspase-1 (Ac-YVAD-cmk), but not a control peptide, was able to inhibit substantially the migration of LC.[52]

Given the critical role of IL-1β in the mobilization of LC it comes as no surprise that in the majority of investigations it has been found that in the absence of this cytokine the acquisition of contact sensitization is sub-optimal.[49,52,53]

An interesting, and only recently considered, aspect of allergen-induced responses in the skin is the influence of age. Recent investigations in mice have revealed that compared with their younger counterparts, older mice (6 months old) display a significantly reduced density of epidermal LC. More importantly, fewer epidermal LC in older mice were mobilized in response to either topical sensitization or intradermal exposure to TNF-α. In contrast, the percentage of epidermal LC that migrated in response to IL-1β was found to be comparable for young and old mice.[54] These results suggest that with increasing age there is a reduction in both the number of LC resident within the epidermis and the responsiveness of these cells to certain stimuli. The fact that LC from older mice were fully responsive to IL-1β suggests that there may in fact be an age-related deficit of cutaneous IL-1β. In parallel investigations it has been found that in humans also there are changes in LC numbers and activity with increasing age. Compared with young adult volunteers (mean age of 23 years), more elderly subjects (mean age 76 years) exhibited a significant reduction in the frequency of epidermal LC, associated with an impaired responsiveness to TNF-α.[35] The relevance of age-related changes in LC biology for senescence of the immune system and disorders of the elderly in general, and for the natural history of allergic contact dermatitis in particular, remains to be clarified.

A two cytokine model of migration wherein induced and up-regulated TNF-α and IL-1β supply independent signals to LC is attractive, but probably represents an over-simplification of the relevant molecular processes. Although there is no doubt that in most or all circumstances TNF-α and IL-1β are essential for the effective mobilization of LC in response to skin sensitization, there has recently been discovered in mice an obligatory role for IL-18 also. This cytokine was characterized originally as an inducer of interferon (IFN)-γ, but has since then been found to contribute in several ways to immune and inflammatory responses.[55-57] Interleukin 18 is expressed by DC (including LC) and keratinocytes, and such expression may be elevated in response to contact allergen.[58-61] Finally, and of relevance to the subject of this article, IL-1β and IL-18 are structurally similar and, although they signal through separate receptors, they use similar intracellular signalling pathways.[55,56]

The first observation was that IL-18, in common with both IL-1β and TNF-α, when administered intradermally to mice is able to induce LC migration and the subsequent accumulation of DC in draining lymph nodes.[62] Importantly, studies performed using neutralizing anti-IL-18 antibody revealed that both LC migration and DC accumulation in response to skin sensitization were dependent upon the availability of IL-18.[62] As IL-18 appears essential for the mobilization of LC following topical sensitization then the intriguing question is where it is sited in the cytokine signalling cascade. Pertinent to this question were the observations that migration induced by exogenous IL-18 is dependent upon intact TNF-α and IL-1β signal transduction pathways. It was found that neutralizing anti-TNF-α antibody inhibited completely both LC migration and DC accumulation stimulated by local injection of IL-18. Moreover, IL-18-induced changes were similarly inhibited by blocking IL-1β signalling pathways with an anti-IL-1RI antibody. Finally, experiments were performed using caspase-1 deficient mice and their wild-type controls. Although in intact animals both IL-1β and IL-18 were able to elicit LC migration, in the knockout mice IL-18 was without effect. The conclusion drawn from these data is that IL-18 is able to cause LC mobilization, but does so in an IL-1β- and TNF-α-dependent manner. The inference is that IL-18 acts upstream of IL-1β and TNF-α, possibly by providing a signal for the up-regulated expression and/or bioavailability of one or other or both of these cytokines.[62] Comparable interpretations can be drawn from studies of IL-18 knockout mice. It was shown, as expected, that in IL-18 deficient mice topical exposure to contact allergens failed to provoke LC migration. Of considerable interest was the observation that LC could be successfully mobilized in IL-18-deficient mice with IL-1β, TNF-α or with IL-18; results that are compatible with IL-18 playing an essential role in LC activation acting upstream of IL-1β and TNF-α.[63] Consistent with the importance of IL-18 for LC migration are demonstrations that in mice treated with neutralizing anti-IL-18 antibody, and in IL-18 knockout mice, there is a significant inhibition of contact hypersensitivity responses.[63,64]

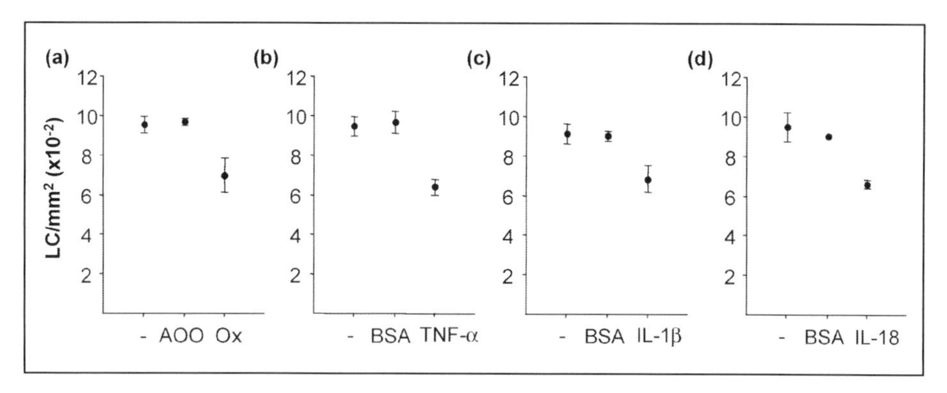

Figure 1. Induction of LC migration by allergen and cytokines. a) Groups of mice (n=2) were exposed topically to 0.5% oxazolone (Ox) suspended in acetone:olive oil (4:1; AOO), to AOO alone, or were left untreated (-). The frequencies of LC were determined 4hr later using indirect immunofluorescence staining of epidermal sheets for MHC class II antigen expression. b-d) Groups of mice (n=2) received intradermal injections (50ng; 30μl) into both ear pinnae of recombinant homologous cytokine suspended in phosphate buffered saline (PBS) containing 0.1% bovine serum albumin (0.1% BSA/PBS). Control mice received injections of carrier protein (0.1% BSA/PBS) alone or were untreated (-). MHC class II$^+$ LC frequencies in epidermal sheets were determined 30min following intradermal administration of TNF-α, or 4hr following treatment with IL-1β or IL-18. Results are expressed as mean LC values/mm^2 (± SE) derived from examination of 10 fields/sample for each of 4 samples.

In summary, evidence available to date indicates that the activation and mobilization of LC in response to skin sensitization requires the coordinated action of at least three cytokines, IL-1β, TNF-α and IL-18, each of which delivers discrete signals to LC and/or keratinocytes. In Figure 1 are displayed representative data demonstrating the ability of a contact allergen (oxazolone), TNF-α, IL-1β and IL-18 to each stimulate LC migration in mice. One model for the cascade of cytokine signalling necessary for LC mobilization is illustrated in Figure 2.

Counter-Regulation by Cytokines

The skin, and in particular the epidermis, is a complex immunological tissue. Epidermal cells (both LC and keratinocytes) express constitutively, or can be stimulated to express, a wide variety of cytokines and chemokines.[21] It is inconceivable that the three cytokines described above which are required for LC migration are not subject to temporal and spatial constraints and the inhibitory activity of other cytokines and cofactors. Presumably, in any given set of circumstances, the vigor and longevity of induced LC migration will be dependent upon the relative availability of promotional and inhibitory stimuli.

An example of a possible counter-regulatory cytokine is interleukin 10 (IL-10), a product of epidermal cells that is up-regulated during skin sensitization,[65] and which is known to inhibit cutaneous inflammatory responses.[66] It was reported by Wang et al[67] that, compared with wild-type controls, IL-10 knockout mice exhibit an increased number of antigen-bearing DC in draining lymph nodes following topical sensitization. The data suggest that the assumed enhancement of LC migration in IL-10-deficient mice is due to increased expression or activity of cytokines that promote mobilization. Thus, there was in IL-10 knockout mice a more marked up-regulation of epidermal mRNA for TNF-α and IL-1β following allergen exposure, and the elevated levels of DC accumulation in draining lymph nodes were inhibited by either neutralizing anti-TNF-α antibody or IL-1 receptor antagonist (IL-1Ra).[67] One interpretation is, therefore, that epidermal IL-10 provides a homeostatic counter-balance for LC mobilization by regulating the level of expression of the promotional cytokines IL-1β and TNF-α.

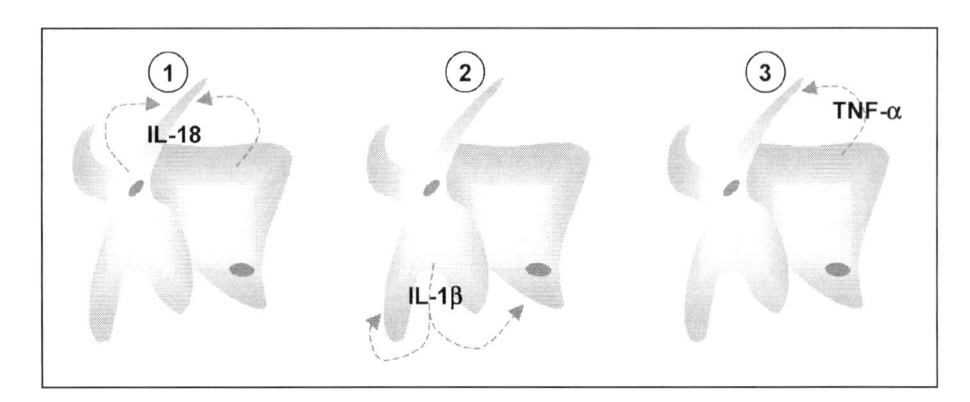

Figure 2. Cytokine signalling for LC migration following skin sensitization. One proposed model for the cytokine signalling cascade required for the mobilization of LC in response to chemical allergen: 1) Following exposure of mice to allergen, LC and/or keratinocytes are stimulated to release bioactive IL-18. Although this cytokine is essential for the initiation of LC migration, IL-18 may also influence directly keratinocyte activation (not shown here). 2) LC produce IL-1β that acts in autocrine fashion on LC (via IL-1RI receptors) to provide one stimulus for migration. IL-1β also interacts with keratinocytes to 3) induce the production of TNF-α, which provides a second independent signal acting on LC in paracrine fashion via TNF-R2 receptors.

A second candidate is transforming growth factor (TGF)-β1, a cytokine known to have pleiotropic properties. The case for active regulation by TGF-β1 is built on several independent lines of evidence. It was found, for instance, that TGF-β1 knockout mice lack epidermal LC.[68] In addition, there are indications that TGF-β1 is able in vitro to modify the response of LC to TNF-α and IL-1,[69] to inhibit the up-regulation by TNF-α of the CCR7 chemokine receptor on DC (CCR7 being known to play an important role in the homing of mobilized LC; see later),[70] and increase the expression by DC of E-cadherin (an adhesion molecule required for retention of LC within the epidermal matrix; see below).[71,72]

Finally, as indicated above, it has been demonstrated, in both man and mouse, that LF is able to inhibit allergen- and IL-1β-induced LC migration; an effect that is almost certainly a function of the ability of LF to inhibit the de novo production by skin cells of TNF-α.[34,37,38,51] LF is an iron-binding protein that is found in exocrine secretions and that is expressed in healthy skin. It is possible that LF provides a molecular mechanism of regulating proinflammatory cytokine responses, and one consequence of this in the skin would be to impose some constraint on cytokine-induced LC activation.

Functional Heterogeneity Among Epidermal LC

Whatever the outcome of the opposing promotional and inhibitory influences of various cytokines and other factors, it is clear that, irrespective of the stimulus that initiates mobilization, only a proportion of local LC (usually no more than about 30%) migrate from the epidermis. This suggests the existence of some functional heterogeneity among LC; the possibility being that this is in turn a function of the length of residence within the epidermal matrix. Indeed there is some evidence for variations among human epidermal LC with respect to their expression of MHC class II.[73] In theory at least, differential mobilization could be attributable to variations in expression of receptors for IL-1β, TNF-α and/or IL-18, and possibly of relevant chemokine receptors. There is, however, one other intriguing possibility based upon the availability of bioactive IL-1β (and possibly other cytokine mediators). There is now evidence that p-glycoprotein (or multidrug resistance gene product 1; MDR-1) may be involved in LC migration. MDR-1 is a plasma membrane protein that serves as an adenosine triphosphate

(ATP)-dependent drug efflux pump, and which is able therefore to confer on cells multidrug resistance. The seminal observation was that neutralizing anti-MDR-1 antibodies, or an antagonist of MDR-1 (verapamil), could inhibit the movement of DC from skin explants and cause the retention of LC within the epidermis.[74] The argument is that if MDR-1 activity is required for the export of IL-1β from cells then presumably this cytokine would be released preferentially by cells that lack membrane ATPase activity. It is interesting therefore, that although many LC have detectable ATPase, not all do.[75] In fact, the proportion of LC that lack ATPase activity (approximately 20%) is not dissimilar to the fraction of epidermal LC that migrate in response to a stimulus. Furthermore, LC that have ATPase will be better protected from the effects of extracellular ATP which is likely to be generated as a result of dermal trauma. In macrophages it has been shown that extracellular ATP is able to cause the post-translational modification of pro-IL-1β and release of the bioactive cytokine by a process mediated via the activation of purinergic P2Z receptors.[76-78] LC express these receptors,[79] so the implication is that cells lacking membrane ATPase will be those that are able to respond to extracellular ATP with the release of active IL-1β. Taken together, the (as yet unproven) hypothesis is that the selection of LC for mobilization in response to an appropriate stimulus is a function of differential cellular ATPase activity.

It should be noted here that another member of the multidrug resistance family (multidrug resistance associated protein 1; MRP1), a transporter of leukotriene C_4 (LTC_4) is also expressed by LC and is necessary for their successful migration.[80] Mobilization from the epidermis and trafficking to lymphatics was markedly impaired in mice lacking MRP1, but activity could be restored with exogenous LTC_4. The conclusion drawn was that MRP1 controls migration to lymph nodes by transporting LTC_4, that in turn allows LC to respond to one of the chemokine ligands (ELC; CCL19) for CCR7 that guides cells from the skin to lymph nodes[80] (see later section on chemokines).

Adhesion Molecules and the Extracellular Matrix

Among the effects induced in LC by promotional cytokines are changes in the expression of those adhesion molecules that permit egress from the epidermis and movement within and through the relevant extracellular matrices. The first step in migration is the need for LC to disassociate themselves from surrounding keratinocytes in the epidermis. The key to this detachment is altered expression of E-cadherin, a calcium-dependent homophilic adhesion molecule. E-cadherin is expressed in normal epidermis by both LC and keratinocytes and has been regarded as the 'glue' that ensures contact between these cells and that retains LC in the skin.[81,82] Mobilization of LC is characterized by the loss of E-cadherin; DC that arrive in the draining lymph nodes displaying little or no expression.[83,84] It has now been established that changes in the expression of E-cadherin by LC following activation are effected by TNF-α, and possibly other proinflammatory cytokines.[85,86] It is likely that in normal circumstances TNF-α is the predominant mediator of changes in E-cadherin expression. As described above, intradermal exposure of mice to TNF-α results in LC migration; a response that can be inhibited by prior exposure of animals to a neutralizing anti-IL-1β antibody. Although in mice treated with anti-IL-1β local exposure to TNF-α fails to stimulate a reduction in their frequency within the epidermis, there are changes in LC morphology. A proportion of LC was found to exhibit a rounded appearance compared with the dendritic morphology of neighbouring cells, or with LC in epidermal sheets prepared from control mice. The LC that displayed such altered morphology were loose and detached from surrounding keratinocytes. The conclusion drawn is that exogenous TNF-α is in this situation providing a signal for LC to disassociate from adjacent keratinocytes (presumably as a consequence of reduced E-cadherin expression), but that in the absence of a second signal from IL-1β LC are unable to continue further on their journey.[48]

That journey, following detachment from other epidermal cells, requires that LC interact with, and traverse, the basement membrane and then gain access to afferent lymphatics by trafficking across the dermal extracellular matrix. A variety of adhesion molecules has been implicated in this process and two of these (α6 integrin and CD44) are considered briefly here.

As mentioned above, a critical step in the migration of LC away from the skin is interaction with, and passage across, the basement membrane at the dermal-epidermal junction. This necessitates the interaction of LC with laminin and this is facilitated by their expression of α6 integrins. Studies in mice revealed that the majority of epidermal LC express the α6 integrin subunit, whereas lymph node DC do not.[87] An anti-α6 integrin antibody known to block binding to laminin was found to inhibit completely the spontaneous migration of LC from skin explants in vitro. Furthermore, this antibody was able in vivo to prevent the migration of LC away from the epidermis induced by TNF-α, and also to impair substantially the accumulation of DC in draining lymph nodes following skin sensitization. An interesting additional observation was that when TNF-α-induced migration was inhibited using anti-α6 antibody many of the LC displayed a rounded morphology and appeared to be detached from neighbouring keratinocytes. These data serve to emphasise again the importance of TNF-α for the initial disassociation of LC from the epidermal matrix, and suggest that one action of IL-1β may be to maintain levels of expression of α6 integrin and/or the β integrin subunits with which α6 dimerises for adhesion to laminin.[87]

The second molecule that warrants consideration is CD44, a transmembrane glycoprotein expressed on the surface of leukocytes, fibroblasts and epidermal cells. Its principal ligand is hyaluronic acid, an important component of extracellular matrices and mediator of cellular interactions.[88,89] Although it has been shown that LC and lymph node DC display comparable levels of CD44,[84] there are other reports demonstrating that epidermal cytokines may have important regulatory influences on the expression of this molecule by LC. Studies performed in vitro with cultured LC revealed that TNF-α and IL-10 have reciprocal antagonistic effects on CD44; the former causing an up-regulation, and the latter a down-regulation. When LC were exposed to both cytokines there was no net change in CD44 levels.[90] A role for CD44 in LC migration was suggested by Gabrilovich et al,[91] who found that inhibition of DC accumulation in draining lymph nodes (and of contact sensitization) caused by retroviral infection of mice was associated with a reduced expression of CD44 (and intercellular adhesion molecule-1; ICAM-1 [CD54]). However, in supplementary investigations the observation was made that contact sensitization could be restored in such mice with IL-12, but in the absence of increased CD44 expression.[92] It may be, however, that these somewhat conflicting data regarding the relevance of CD44 for LC migration reflect the fact that it is certain isoforms of the adhesion molecule that are most important. It has been demonstrated that incubation of cultured DC with TNF-α causes, among other changes, the appearance of CD44-v9, a CD44 exon 9 splice variant.[93] This isoform of CD44 has been shown to confer metastatic potential on carcinoma cells,[94] and the argument is that it might also play a role in the migration of LC. There is evidence to support this as antibodies specific for certain CD44 isoforms have been found to inhibit LC migration.[95] On the basis of these data it is likely that cytokine activation of LC is associated with the induced or elevated expression of certain isoforms of CD44 that are required for the effective trafficking of cells through the hyaluronate-rich dermis.

Before leaving consideration of the nature of interactions between LC and the extracellular matrix, it is relevant to acknowledge that passage across the basement membrane will require proteolysis. It is now established that LC possess matrix metalloproteinase (MMP)-9,[96,97] and that the activity of this enzyme is up-regulated following skin sensitization, possibly in response to IL-1β and/or TNF-α.[98] More recently, DC from the skin have been shown to express MMP-2 also.[99] Importantly, blocking antibodies specific for MMP-9 and MMP-2, and various MMP inhibitors have been shown to affect LC migration.[99-101] The importance of

MMP function for the effective acquisition of skin sensitization is illustrated by investigations of mice deficient in MMP-3 (stromelysin 1) which were found to have a profound impairment of contact hypersensitivity.[102]

In summary, it is apparent that the effective migration of LC following skin sensitization requires the local availability of certain epidermal cytokines that collectively are responsible for effecting a variety of changes, including those that permit escape from the epidermis, movement through the extracellular matrix and access to afferent lymphatics. However, superimposed upon these events appears to be the need for directional guidance and this is provided by chemokine-chemokine receptor interactions.

Chemokines, Langerhans Cell Migration and Homing to Lymph Nodes

Chemokines are a superfamily of proteins that influence immune and inflammatory responses, primarily by regulating the movement and localization of leukocytes.[103-105] In recent years there has been a growing interest in the roles played by chemokines in the orchestration of DC function and in the migration of LC. In the context of the latter, changes in chemokine responsiveness of LC appear to be required for both allowing movement away from the skin and the subsequent homing of cells to the paracortical regions of draining lymph nodes.

Certain chemokines are considered to be inflammatory, insofar as they attract cells to sites of inflammation. It has been proposed that some of these may also facilitate the recruitment of immature cells from the blood to replace those LC that have been stimulated to migrate from the skin. Whether or not this is the case, it is clear that the induced maturation of DC is associated with a rapid reduction in the level of expression of several chemokine receptors, including CCR1 and CCR5, that confer responsiveness to chemokines such as macrophage inflammatory protein (MIP) 1α and RANTES (Regulated upon Activation, normal T cell Expressed and Secreted).[106,107] There is evidence that these changes in chemokine receptor expression can be effected by TNF-α and IL-1.[106] Taken together these observations suggest a coordinated interplay in the epidermis between proinflammatory cytokines and the expression of certain chemokine receptors that will not only stimulate the migration of responsive LC away from the skin, but also encourage their replacement by immature precursors recruited from the blood. The local trauma that is associated with skin sensitization will result in the up-regulated expression of various inflammatory chemokines that will attract immature cells into the skin. However, the action of IL-1β and TNF-α on resident activated LC will down-regulate responsiveness of the cells to such chemokines and ensure that their departure from the epidermis is not impeded.

During maturation there are, concomitant with down-regulation of CCR1 and CCR5, changes in the expression of CCR7 by LC. This chemokine receptor has as its ligands secondary lymphoid tissue chemokine (SLC; CCL21) and EBV-induced molecule 1 ligand (ELC; CCL19). The former is produced by high endothelial venules in peripheral lymph nodes, and to a lesser extent by stromal cells within the lymph node paracortex, whereas the latter is produced by DC and stromal cells in the paracortex.[108-111] The CCR7 receptor is absent from resting epidermal LC, but is up-regulated during DC maturation, at least one stimulus for increased expression being TNF-α.[106,107,112-114] The role of CCR7-ligand interactions has been clarified further by investigation of mutant mice (paucity of lymph node T cells; *plt* mutation) that lack lymphoid SLC and exhibit defective T cell homing to lymph nodes. Such mice display a markedly reduced accumulation of antigen-bearing DC in draining lymph nodes following skin sensitization.[115] In fact, there is now evidence that there exists a second gene that encodes for SLC that, unlike the gene deleted in *plt* mice, is expressed in non-lymphoid tissue and may influence the interaction of migrating cells with the lymphatic endothelium.[116] Notwithstanding the complication of gene duplication of SLC chemokines, definitive evidence that ligand interactions with CCR7 are essential for the migration of LC to draining lymph nodes derived from investigations using CCR7 gene-targeted mice in which this process is completely lost.[117]

The combined effect of these chemokine receptor-chemokine interactions is to allow LC to leave the skin (through down-regulation of CCR1 and CCR5) in the face of local trauma and inflammation. Subsequently, as a consequence of up-regulated CCR7 expression, the emigrating LC are first directed to afferent lymphatics and then positioned within the appropriate compartment of the paracortex. With regard to the organization of cells arriving in the lymph node it is worth noting that an important cellular source of ELC (CCL19; one of the CCR7 ligands) are DC in lymph nodes. The consequence is that as migrating cells follow the CCR7-ELC chemotactic gradient they are likely to be positioned adjacent to other DC already resident in draining lymph nodes. This may in turn facilitate cross-talk and/or antigen transfer between DC as part of the antigen presentation process.

Concluding Comments

The interplay between cytokines and chemokines, and the changes they effect in the phenotype of LC, are both complex and biologically elegant. The end result is that in response to cytokines that signal potential 'danger' at skin surfaces, and guided by chemotactic gradients, LC move from the site of insult to lymph nodes where cutaneous immune responses are initiated. Many important questions remain to be addressed. For instance, it is well established that skin trauma that is not (at least initially) associated with antigenic exposure is able to stimulate LC migration and the accumulation of DC in draining lymph nodes. What is not clear, however, is whether, in the absence of encounter with antigen, the maturational changes to which migrating cells are subject remain the same. Moreover, it is assumed, but not proven, that all chemical allergens drive LC migration and maturation via identical mechanisms. An intriguing question that needs now to be addressed is whether different forms of chemical allergen induce LC activation and migration through the same or different processes.

References

1. Banchereau J, Steinman RM. Dendritic cells and the control of immunity. Nature 1998; 392:245-252.
2. Pulendran B, Maraskovsky E, Banchereau J et al. Modulating the immune response with dendritic cells and their growth factors. Trends Immunol 2001; 22:41-47.
3. Romani N, Schuler G. The immunologic properties of epidermal Langerhans cells as part of the dendritic cell system. Springer Semin Immunopathol 1992; 13:265-279.
4. Lappin MB, Kimber I, Norval M. The role of dendritic cells in cutaneous immunity. Arch Derm Res 1996; 288:109-121.
5. Cella M, Sallusto F, Lanzavecchia A. Origin, maturation and antigen presenting function of dendritic cells. Curr Opinion Immunol 1997; 9:10-16.
6. Ardavin C, Martinez del Hoyo G, Martin P et al. Origin and differentiation of dendritic cells. Trends Immunol 2001; 22:691-700.
7. Martinez del Hoyo G, Martin P, Hernandez Vargas H et al. Characterization of a common precursor population for dendritic cells. Nature 2002; 415:1043-1047.
8. Cruz PD Jr, Tigelaar RE, Bergstresser PR. Langerhans cells that migrate to skin after intravenous infusion regulate the induction of contact hypersensitivity. J Immunol 1990; 144:2486-2492.
9. Jakob T, Ring J, Udey MC. Multistep navigation of Langerhans/dendritic cells in and out of the skin. J Allergy Clin Immunol 2001; 108:688-696.
10. Macatonia SE, Edwards AJ, Knight SC. Dendritic cells and the initiation of contact sensitivity to fluorescein isothiocyanate. Immunology 1996; 59:509-514.
11. Macatonia SE, Knight SC, Edwards AJ et al. Localization of antigen on lymph node dendritic cells after exposure to the contact sensitizer fluorescein isothiocyanate. Functional and morphological studies. J Exp Med 1987; 166:1654-1667

12. Kripke ML, Munn CG, Jeevan A et al. Evidence that cutaneous antigen-presenting cells migrate to regional lymph nodes during contact sensitization. J Immunol 1990; 145:2833-2838.
13. Cumberbatch M, Kimber I. Phenotypic characteristics of antigen-bearing cells in the draining lymph nodes of contact sensitised mice. Immunology 1990; 71:404-410.
14. Cumberbatch M, Illingworth I, Kimber I. Antigen-bearing cells in the draining lymph nodes of contact sensitised mice: cluster formation with lymphocytes. Immunology 1991; 74:139-145.
15. Kinnaird A, Peters SW, Foster JR et al. Dendritic cell accumulation in the draining lymph nodes during the induction phase of contact allergy in mice. Int Arch Allergy Appl Immunol 1989; 89:202-210.
16. Schuler G, Steinman RM. Murine epidermal Langerhans cells mature into potent immunostimulatory dendritic cells in vitro. J Exp Med 1985; 161:526-546.
17. Streilein JW, Grammer S. In vitro evidence that Langerhans cells can adopt two functionally distinct forms capable of antigen presentation to T lymphocytes. J Immunol 1989; 143:3925-3933.
18. Heufler C, Koch F, Schuler G. Granulocyte/macrophage colony-stimulating factor and interleukin 1 mediate the maturation of murine epidermal Langerhans cells into potent immunostimulatory dendritic cells. J Exp Med 1988; 167:700-705.
19. Henri S, Vremec D, Kamath A et al. The dendritic cell populations of mouse lymph nodes. J Immunol 2001; 167:741-748.
20. Kimber I, Cumberbatch M. Dendritic cells and cutaneous immune responses to chemical allergens. Toxicol Appl Pharmacol 1992; 117:137-146.
21. Kimber I, Cumberbatch M, Dearman RJ et al. Langerhans cell migration and cellular interactions. In: Lotze MT, Thomson AW, eds. Dendritic Cells. Biology and Clinical Applications.San Diego, Academic Press, 1999:295-310.
22. Kimber I, Cumberbatch M, Dearman RJ et al. Cytokines and chemokines in the initiation and regulation of epidermal Langerhans cell mobilization. Br J Dermatol 2000; 142:401-412.
23. Tse Y, Cooper KD. Cutaneous dermal Ia+ cells are capable of initiating delayed-type hypersensitivity responses. J Invest Dermatol 1990; 94:267-272.
24. Bacci S, Alard P, Dai R et al. High and low doses of hapten dictate whether dermal or epidermal antigen-presenting cells promote contact hypersensitivity. Eur J Immunol 1997; 27:442-448.
25. Sato K, Imai Y, Irimura T. Contribution of dermal macrophage trafficking in the sensitization phase of contact hypersensitivity. J Immunol 1998; 161:6835-6844.
26. Moodycliffe AM, Kimber I, Norval M. The effect of ultraviolet B irradiation and urocanic acid isomers on dendritic cell migration. Immunology 1992; 77:394-399.
27. Moodycliffe AM, Kimber I, Norval M. Role of tumour necrosis factor-α in ultraviolet B light-induced dendritic cell migration and suppression of contact hypersensitivity. Immunology 1994; 81:79-84.
28. Cumberbatch M, Scott RC, Basketter DA et al. Influence of sodium lauryl sulphate on 2,4-dinitrochlorobenzene-induced lymph node activation. Toxicology 1993; 77:181-191.
29. Cumberbatch M, Dearman RJ, Groves RW et al. Differential regulation of epidermal Langerhans cell migration by interleukin (IL)-1α and IL-1β during irritant- and allergen-induced cutaneous immune responses. Toxicol Appl Pharmacol (in press).
30. Enk AH, Katz SI. Early molecular events in the induction phase of contact sensitivity. Proc Natl Acad Sci USA 1992; 89:1398-1402.
31. Cumberbatch M, Kimber I. Dermal tumour necrosis factor-α induces dendritic cell migration to draining lymph nodes, and possibly provides one stimulus for Langerhans cell migration. Immunology 1992; 75:257-263.
32. Cumberbatch M, Fielding I, Kimber I. Modulation of epidermal Langerhans cell frequency by tumour necrosis factor-α. Immunology 1994; 81:395-401.
33. Cumberbatch M, Griffiths CEM, Tucker SC et al. Tumour necrosis factor-α induces Langerhans cell migration in humans. Br J Dermatol 1999; 141:192-200.
34. Griffiths CEM, Cumberbatch M, Tucker SC et al. Exogenous topical lactoferrin inhibits allergen-induced Langerhans cell migration and cutaneous inflammation in humans. Br J Dermatol 2001; 144:715-725.
35. Bhushan M, Cumberbatch M, Dearman RJ et al. Tumour necrosis factor-α-induced migration of human Langerhans cells: the influence of ageing. Br J Dermatol 2002; 146:32-40.

36. Cumberbatch M, Kimber I. Tumour necrosis factor-α is required for accumulation of dendritic cells in draining lymph nodes and for optimal contact sensitization. Immunology 1995; 84:31-35.
37. Kimber I, Cumberbatch M, Dearman RJ et al. Lactoferrin: influences on Langerhans cells, epidermal cytokines, and cutaneous inflammation. Biochem Cell Biol 2002; 80:103-107.
38. Cumberbatch M, Dearman RJ, Uribe-Luna S et al. Regulation of epidermal Langerhans cell migration by lactoferrin. Immunology 2000; 100:21-28.
39. Cumberbatch M, Dearman R, Kimber I. Inhibition by dexamethasone of Langerhans cell migration: influence of epidermal cytokine signals. Immunopharmacol 1999; 41:235-243.
40. Wang B, Kondo S, Shivji GM et al. Tumour necrosis factor receptor II (p75) signalling is required for the migration of Langerhans cells. Immunology 1996; 88:284-288.
41. Wang B, Fujisawa H, Zhuang L et al. Depressed Langerhans cell migration and reduced contact hypersensitivity response in mice lacking TNF receptor p75. J Immunol 1997; 159:6148-6155.
42. Koch F, Heufler C, Kampgen E et al. Tumor necrosis factor α maintains the viability of murine epidermal Langerhans cells in culture but in contrast to granulocyte/macrophage colony-stimulating factor, does not induce their functional maturation. J Exp Med 1990; 171:159-172.
43. Ryffel B, Brockhaus M, Greiner B et al. Tumour necrosis factor receptor distribution in human lymphoid tissue. Immunology 1991; 74:446-452.
44. Larregina A, Morelli A, Kolkowski E et al. Flow cytometric analysis of cytokine receptors on human Langerhans cells. Changes observed after short term culture. Immunology 1996; 87:317-325.
45. Cumberbatch M, Dearman RJ, Kimber I. Interleukin 1β and the stimulation of Langerhans cell migration: comparisons with tumour necrosis factor α. Arch Dermatol Res 1997; 289:277-284.
46. Cumberbatch M, Dearman RJ, Kimber I. Characteristics and regulation of the expression of interleukin 1 receptors by murine Langerhans cells and keratinocytes. Arch Dermatol Res 1998; 290:688-695.
47. Cumberbatch M, Dearman RJ, Kimber I. Langerhans cell migration in mice requires intact type I interleukin 1 receptor (IL-1RI) function. Arch Dermatol Res 1999; 291:357-361.
48. Cumberbatch M, Dearman RJ, Kimber I. Langerhans cells require signals from both tumour necrosis factor-α and interleukin-1β for migration. Immunology 1997; 92:388-395.
49. Enk AH, Angeloni VL, Udey MC et al. An essential role for Langerhans cell-derived IL-1β in the initiation of primary immune responses in the skin. J Immunol 1993; 150:3698-3704.
50. Kimber I, Dearman RJ, Cumberbatch M et al. Langerhans cells and chemical allergy. Curr Opinion Immunol 1998; 10:614-619.
51. Cumberbatch M, Bhushan M, Dearman RJ et al. IL-1β-induced Langerhans cell migration and TNF-α production in human skin: regulation by lactoferrin. Cliln Exp Immunol 2003; 132:352-359.
52. Antonopoulos C, Cumberbatch M, Dearman RJ et al. Functional caspase-1 is required for Langerhans cell migration and optimal contact sensitization in mice. J Immunol 2001; 166:3672-3677.
53. Shornick LP, De Togni P, Mariathasan S et al. Mice deficient in IL-1β manifest impaired contact hypersensitivity to trinitrochlorobenzene. J Exp Med 1996; 183:1427-1436.
54. Cumberbatch M, Dearman RJ, Kimber I. Influence of ageing on Langerhans cell migration in mice: identification of a putative deficiency of epidermal interleukin-1β. Immunology 2002; 105:466-477.
55. Dinarello CA, Novick D, Puren AJ et al. Overview of interleukin-18: more than an interferon-γ inducing factor. J Leukocyte Biol 1998; 63:658-664.
56. Akira S. The role of IL-18 in innate immunity. Curr Opinion Immunol 2000; 12:59-63.
57. McInnes IB, Gracie A, Leung BP et al. Interleukin 18: a pleiotropic participant in chronic inflammation. Immunol Today 2000; 21:312-315.
58. Stoll S, Muller G, Kurimoto M et al. Production of IL-18 (IFN-γ-inducing factor) messenger RNA and functional protein by murine keratinocytes. J Immunol 1997; 159:298-302.
59. Stoll S, Jonuleit H, Schmitt E et al. Production of functional IL-18 by different subtypes of murine and human dendritic cells (DC): DC-derived IL-18 enhances IL-12-dependent Th1 development. Eur J Immunol 1998; 28:3231-3239.
60. Naik SM, Cannon G, Burbach GJ et al. Human keratinocytes constitutively express interleukin-18 and secrete biologically active interleukin-18 after treatment with pro-inflammatory mediators and dinitrochlorobenzene. J Invest Dermatol 1999; 113:766-772.

61. Mee JB, Alam Y, Groves RW. Human keratinocytes constitutively produce interleukin-18. Br J Dermatol 2000; 143:330-336.
62. Cumberbatch M, Dearman RJ, Antonopoulos C et al. Interleukin (IL)-18 induces Langerhans cell migration by a tumour necrosis factor-α- and IL-1β-dependent mechanism. Immunology 2001; 102:323-330.
63. Antonopoulos C, Cumberbatch M, Dearman RJ et al. IL-18 is a key proximal mediator of contact hypersensitivity and allergen-induced Langerhans cell migration in murine epidermis. Submitted for publication.
64. Wang B, Feliciani C, Howell BG et al. Contribution of Langerhans cell-derived IL-18 to contact hypersensitivity. J Immunol 2002; 168:3303-3308.
65. Enk AH, Katz SI. Identification and induction of keratinocyte-derived IL-10. J Immunol 1992; 149:92-95.
66. Berg DJ, Leach MW, Kuhn R et al. Interleukin 10 but not interleukin 4 is a natural suppressant of cutaneous inflammatory responses. J Exp Med 1995; 182:99-108.
67. Wang B, Zhuang L, Fujisawa et al. Enhanced epidermal Langerhans cell migration in IL-10 knockout mice. J Immunol 1999; 162:277-283.
68. Borkowski TA, Letterio JJ, Farr AG et al. A role for endogenous transforming growth factor β1 in Langerhans cell biology: the skin of transforming growth factor β1 null mice is devoid of epidermal Langerhans cells. J Exp Med 1996; 184:2417-2422.
69. Geissmann F, Revy P, Regnault A et al. TGF-β1 prevents the noncognate maturation of human dendritic Langerhans cells. J Immunol 199; 162:4567-4575.
70. Sato K, Kawaski H, Nagayama H et al. TGF-beta 1 reciprocally controls chemotaxis of human peripheral blood monocyte-derived dendritic cells via chemokine receptors. J Immunol 2000; 164:2285-2295.
71. Geissmann F, Prost C, Monnet JP et al. Transforming growth factor beta 1, in the presence of granulocyte/macrophage colony-stimulating factor and interleukin 4, induces differentiation of human peripheral blood monocytes into dendritic Langerhans cells. J Exp Med 1998; 187:961-966.
72. Riedl E, Stockl J, Majdic O et al. Functional involvement of E-cadherin in TGF-beta 1-induced cell cluster formation of in vitro developing Langerhans cell-type dendritic cells. J Immunol 2000; 165:1381-1386.
73. Shibaki A, Meunier L, Ra C et al. Differential responsiveness of Langerhans cell subsets of varying phenotypic states in normal human epidermis. J Invest Dermatol 1995; 104:42-46.
74. Randolph GJ, Beaulieu S, Pope M et al. A physiologic function for p-glycoprotein (MDR-1) during the migration of dendritic cells from the skin via afferent lymphatic vessels. Proc Natl Acad Sci USA 1998; 95:6924-6929.
75. Girolomoni G, Santantonio ML, Pastore S et al. Epidermal Langerhans cells are resistant to the permeabilizing effects of extracellular ATP: in vitro evidence supporting a protective role of membrane ATPase. J Invest Dermatol 1993; 100:282-287.
76. Perregauz D, Gabel CA. Interleukin 1β maturation and release in response to ATP and nigericin. J Biol Chem 1994; 269:15195-15203.
77. Griffiths RJ, Stam EJ, Downs JT et al. ATP induces the release of IL-1 from LPS-primed cells in vivo. J Immunol 1997; 154:2821-2828.
78. Ferrari D, Chiozzi P, Falzoni S et al. Extracellular ATP triggers IL-1β release by activating the purinergic receptor of human macrophages. J Immunol 1997; 159:1451-1458.
79. Di Virgilio F. The P2Z purinoreceptor: an intriguing role in immunity, inflammation and cell death. Immunol Today 1995; 16:524-528.
80. Robbiani DF, Finch RA, Jager D et al. The leukotriene C(4) transporter MRP1 regulates CCL19 (MIP-3beta, ELC)-dependent mobilization of dendritic cells to lymph nodes. Cell 2000;103:757-768.
81. Tang A, Amagai M, Granger LG et al. Adhesion of epidermal Langerhans cells to keratinocytes mediated by E-cadherin. Nature 1993; 361:82-85.
82. Blauvelt A, Katz SI, Udey MC. Human Langerhans cells express E-cadherin. J Invest Dermatol 1995; 104:293-296.
83. Borkowski TA, van Dyke BJ, Schwarzenberger K et al. Expression of E-cadherin by murine dendritic cells. E-cadherin as a dendritic cell differentiation antigen characteristic of epidermal Langerhans cells and related cells. Eur J Immunol 1994; 24:2767-2774.

84. Cumberbatch M, Dearman RJ, Kimber I. Adhesion molecule expression by epidermal Langerhans cells and lymph node dendritic cells: a comparison. Arch Dermatol Res 1996; 288:739-744.
85. Schwarzenberger K, Udey MC. Contact allergens and epidermal proinflammatory cytokines modulate Langerhans cell E-cadherin expression in situ. J Invest Dermatol 1996; 106:553-558.
86. Jakob T, Udey MC. Regulation of E-cadherin-mediated adhesion in Langerhans cell-like dendritic cells by inflammatory mediators that mobilize Langerhans cells in vivo. J Immunol 1998; 160:4067-4073.
87. Price AA, Cumberbatch M, Kimber I et al. α6 integrins are required for Langerhans cell migration from the epidermis. J Exp Med 1997; 186:1725-1725.
88. Miyake K, Underhill CB, Lesley J et al. Hyaluronate can function as a cell adhesion molecule and CD44 participates in hyaluronate recognition. J Exp Med 1990; 172:69-75.
89. Hogg N, Landis RC. Adhesion molecules in cell interactions. Curr Opinion Immunol 1993; 5:383-390.
90. Osada A, Nakashima H, Furue et al. Up-regulation of CD44 expression by tumor necrosis factor-α is neutralized by interleukin-10 in Langerhans cells. J Invest Dermatol 1995; 105:124-127.
91. Gabrilovich DI, Woods GM, Patterson S et al. Retrovirus-induced immunosuppression via blocking of dendritic cell migration and down-regulation of adhesion molecules. Immunology 1994; 82:82-87.
92. Williams NJ, Harvey JJ, Duncan I. IL-12 restores dendritic cell function and cell mediated immunity in retrovirus infected mice. Cell Immunol 1998; 183:121-130.
93. Sallusto F, Lanzavecchia A. Efficient presentation of soluble antigen by cultured human dendritic cells is maintained by granulocyte/macrophage colony-stimulating factor plus interleukin 4 and is downregulated by tumor necrosis factor α. J Exp Med 1994; 179:1109-1118.
94. Gunthert U, Hofmann M, Rudy W et al. A new variant of glycoprotein CD44 confers metastatic potential to rat carcinoma cells. Cell 1991; 65:13-24.
95. Weiss JM, Sleeman J, Renkl AC et al. An essential role for CD44 variant isoforms in epidermal Langerhans cell and blood dendritic cell function. J Cell Biol 1997; 137:1137-1147.
96. Kobayashi Y. Langerhans cells produce type IV collagenase (MMP-9) following epicutaneous stimulation with haptens. Immunology 1997; 90:496-501.
97. Uchi H, Imayama S, Kobayashi Y et al. Langerhans cells express matrix metalloproteinase-9 in the human epidermis. J Invest Dermatol 1998; 111:1232-1233.
98. Saren P, Welgus HG, Kovanen PT. TNF-α and IL-1β selectively induce expression of 92-kDa gelatinase by human macrophages. J Immunol 1996; 157:4149-4165.
99. Ratzinger G, Stoitzner P, Ebner S et al. Matrix metalloproteinase 9 and 2 are necessary for the migration of Langerhans cells and dermal dendritic cells from human and murine skin. J Immunol 2002; 168:4361-4371.
100. Kobayashi Y, Matsumoto M, Kotani et al. Possible involvement of matrix metalloproteinase-9 in Langerhans cell migration and maturation. J Immunol 1999; 163:5989-5993.
101. Lebre MC, Kalinski P, Das PK et al. Inhibition of contact sensitizer-induced migration of human Langerhans cells by matrix metalloproteinase inhibitors. Arch Dermatol Res 1999; 291:447-452.
102. Wang M, Qin X, Mudgett JS et al. Matrix metalloproteinase deficiencies affect contact hypersensitivity: stromelysin-1 deficiency prevents the response and gelatinase B deficiency prolongs the response. Proc Natl Acad Sci USA 1999; 96:6885-6889.
103. Rollins B. Chemokines. Blood 1997; 90:909-928.
104. Mantovani A. The chemokine system: redundancy for robust outputs. Immunol Today 1999; 20:254-257.
105. Zlotnik A, Morales J, Hedrick JA. Recent advances in chemokines and chemokine receptors. Crit Rev Immunol 1999; 19:1-47.
106. Sozzani S, Allavena P, D'Amico G et al. Differential regulation of chemokine receptors during dendritic cell maturation: a model for their trafficking properties. J Immunol 1998; 161:1083-1086.
107. Sallusto F, Schaerli P, Loetscher P et al. Rapid and coordinated switch in chemokine receptor expression during dendritic cell maturation. Eur J Immunol 1998; 28:2760-2769.
108. Gunn MD, Tangemann K, Tam C et al. A chemokine expressed in lymphoid high endothelial venules promotes the adhesion and chemotaxis of naïve T lymphocytes. Proc Natl Acad Sci USA 1998; 95:258-263.

109. Hedrick JA, Zlotnik A. Identification and characterization of a novel beta chemokine containing six conserved cysteines. J Immunol 1997; 159:1589-1593
110. Yoshida R, Imai T, Hieshima K et al. Molecular cloning of a novel human CC chemokine EBI1-ligand chemokine that is a specific functional ligand for EBI1, CCR7. J Biol Chem 1997; 272:13803-13809.
111. Ngo VN, Tang HL, Cyster JG. Epstein-Barr virus-induced molecule 1 ligand chemokine is expressed by dendritic cells in lymphoid tissues and strongly attracts naïve T cells and activated B cells. J Exp Med 1998; 188:181-191.
112. Yanagihara S, Komura E, Nagafune J et al. EBI1/CCR7 is a new member of dendritic cell chemokine receptor that is up-regulated upon maturation. J Immunol 1998; 161:3096-3102.
113. Sallusto F, Palermo B, Lenig D et al. Distinct patterns and kinetics of chemokine production regulate dendritic cell function. Eur J Immunol 1999; 29:1617-1625.
114. Saeki H, Moore AM, Brown MJ et al. Secondary lymphoid-tissue chemokine (SLC) and CC chemokine receptor 7 (CCR7) participate in the emigration pathway of mature dendritic cells from the skin to regional lymph nodes. J Immunol 1999; 162:2472-2475.
115. Gunn MD, Kyuwa S, Tam C et al. Mice lacking expression of secondary lymphoid organ chemokine have defects in lymphocyte homing and dendritic cell localization. J Exp Med 1999; 189:451-460.
116. Vassileva G, Soto H, Zlotnik A et al. The reduced expression of 6Ckine in the plt mouse results from the deletion of one of two 6Ckine genes. J Exp Med 1999; 190:1183-1188.
117. Forster R, Schubel A, Breitfeld D et al. CCR7 coordinates the primary immune response by establishing functional microenvironments in secondary lymphoid organs. Cell 1999; 99:23-33.

CHAPTER 4

Contribution of CD4$^+$ and CD8$^+$ T Cells in Contact Hypersensitivity and Allergic Contact Dermatitis

Pierre Saint-Mezard, Frédéric Bérard, Bertrand Dubois, Dominique Kaiserlian and Jean-François Nicolas

Summary

Allergic contact dermatitis (ACD) and contact hypersensitivity (CHS) are delayed-type hypersensitivity reactions which are mediated by hapten specific T cells. During the sensitisation phases, both CD4$^+$ and CD8$^+$ T cell precursors are activated in the draining lymph nodes by presentation of haptenated peptides by skin dendritic cells. Subsequent hapten skin painting induces the recruitment of T cells at the site of challenge which induce inflammatory signals and apoptosis of epidermal cells, leading to the development of a skin inflammatory infiltrate and of clinical symptoms. There have been major controversies on the respective role of CD4$^+$ and CD8$^+$ T cells in the development of the CHS inflammatory reaction. Experimental studies from the last 10 years have demonstrated that, in normal CHS responses to strong haptens, CD8$^+$ type 1 T cells are effector cells of CHS while CD4$^+$ T cells are endowed with down-regulatory functions. The latter may correspond to the recently described CD4$^+$CD25$^+$ regulatory T cell population. However, in some instances, especially those where there is a deficient CD8$^+$ T cell pool, CD4$^+$ T cells can be effector cells of CHS. Ongoing studies will have to confirm that the pathophysiology of human ACD is similar to the mouse CHS and that the CHS response to weak haptens, the most frequently involved in human ACD, is similar to that reported for strong haptens.

Introduction—Delayed-Type Hypersensitivity and Contact Hypersensitivity

Contact hypersensitivity (CHS) is a delayed type hypersensitivity (DTH) reaction, i.e., a skin inflammatory reaction due to the activation of antigen-specific T cells. On the contrary to classical DTH, which needs intradermal injection of exogenous protein, initiation of CHS is generated by topical application on the epidermis of sensitizing chemicals products: nickel, chrome, dinitrofluorobenzene (DNFB), trinitrochlorobenzene (TNCB), oxazolone (OX). CHS is one form of DTH reactions and as such was considered to be CD4$^+$ T cell-mediated. Since cutaneous infiltrates in human allergic contact dermatitis (ACD) show a clear preponderance of CD4$^+$ T cells, it is not surprising that this T cell subset has most often be held responsible for

Immune Mechanisms in Allergic Contact Dermatitis, edited by Andrea Cavani and Giampiero Girolomoni. ©2005 Eurekah.com.

mediating CHS and ACD. However, it has become clear that both $CD4^+$ and $CD8^+$ T cells can act as effector cells in both DTH and CHS reactions.[1-3]

Studies from the last 10 years have emphasized that $CD8^+$ T cells were the main effector cells of CHS while $CD4^+$ T cells behave as down-regulory cells.[4-6] It is noteworthy that there are still some controversies as to whether $CD8^+$ T cells are effector cells of CHS in all strains of mice and for all types of haptens.[7] Similarly, the precise contribution of CD4 and CD8 T cells in human ACD is still not known.[8]

$CD8^+$ T cells are now known to mediate DTH responses in allergic contact dermatitis, drug eruptions, asthma, and autoimmune diseases.[3] This inflammatory effector capability of $CD8^+$ cytotoxic T cells was previously poorly recognized, but there is now considerable evidence that these diseases may be mediated by $CD8^+$ DTH cells. The difference between $CD8^+$ T cells and $CD4^+$ T cells mediating DTH may relate to the molecular mechanisms by which antigens are processed and presented to the T cells. Antigens external to the cell are phagocytosed and processed for presentation on MHC class II molecules (e.g., HLA-DR) to $CD4^+$ T cells. In contrast, internal cytoplasmic antigens are processed by the endogenous pathway for presentation on MHC class I molecules (e.g., HLA-A, -B, and -C) to $CD8^+$ T cells. External allergens can also enter the endogenous pathway to be presented to $CD8^+$ T cells. These include many contact sensitizers, chemical and protein respiratory allergens, viral antigens, metabolic products of drugs, and autoantigens. Haptens are also able to directely interact with peptides which are already in the groove of MHC class II and class I molecules.[4] Thus $CD8^+$ and $CD4^+$ T cells could be activated in the lymph nodes by skin antigen presenting cells (APC) expressing haptenated peptides.

Pathophysiology of CHS—General Scheme

Knowledge of the pathophysiology of ACD is derived chiefly from animal models in which the skin inflammation induced by hapten painting of the skin is referred to as CHS. Two temporally and spacially dissociated phases are usually necessary to achieve optimal CHS reaction: the sensitization and the elicitation phase (Fig. 1). We describe here the well-accepted pathophysiological pathways of CHS and ACD

The sensitization phase (also known as afferent phase) occurs at the first skin contact with a hapten and leads to the priming and expansion of hapten specific T cells in lymph nodes. Topically applied hapten is taken up by skin dendritic cells (DC), especially Langerhans cells (LC), which migrate from the epidermis to the para-cortical area of draining lymph nodes, where they present haptenated peptide/MHC molecule complexes to hapten-specific T cell precursors. Specific T cells emigrate from the lymph nodes and enter the blood through the thoractic duct and recirculate to tissues including the skin.

The elicitation phase occurs a few hours after a subsequent challenge of the skin with the same hapten. Hapten is uptaken by skin APC, particularly skin DC and keratinocytes which present haptenated peptides to specific T cells which patrol in the skin. Activated T cells produce type 1 cytokines interferon (IFN)-γ, activate skin resident cells which produce cytokines and chemokines allowing the recruitment of the polymorphous cellular infiltrate characteristic of CHS. This efferent phase of CHS takes 72 hours in human and 24 to 48 hours in the mouse. The inflammatory reaction persists over several days and progressively decreases upon physiological down-regulating mechanisms which could also be explained by the disappearance of the hapten.

CHS to Strong Haptens

The vast majority of available data on CHS have been obtained with strong sensitizers which have unique chemical and immunological properties:

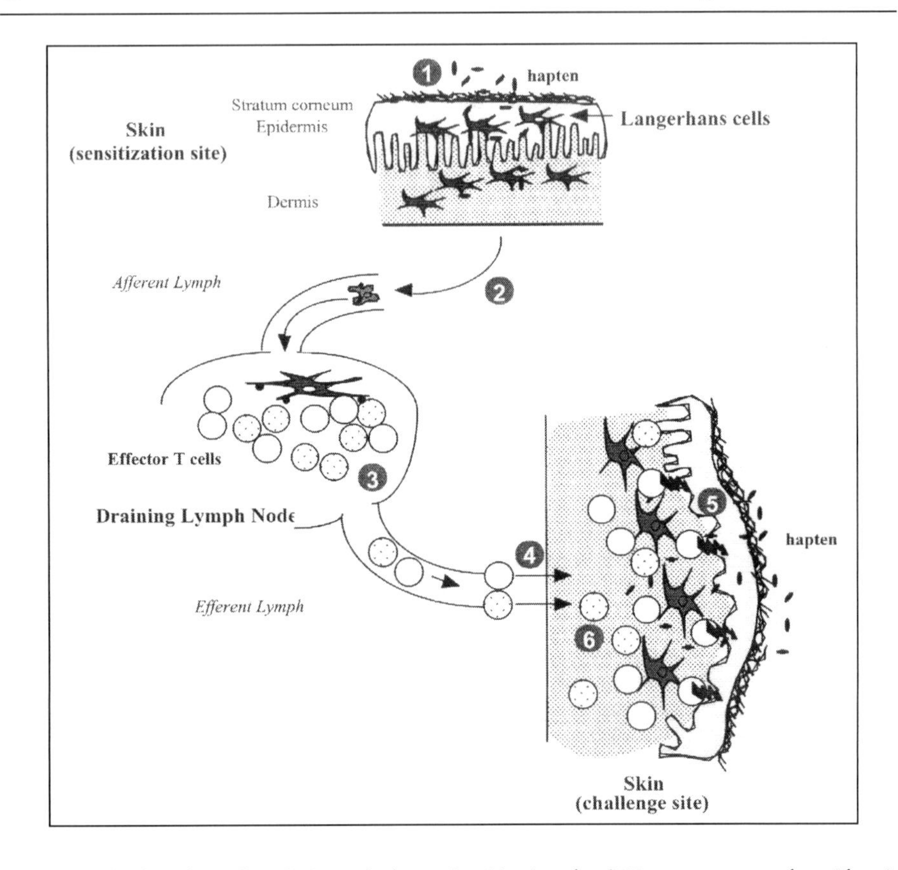

Figure 1. Pathophysiology of CHS. General scheme. Sensitization phase) Haptens penetrate the epidermis (step 1) and are uptaken by skin LC which migrate to draining lymph nodes (step 2) where they present haptenated peptides to T cells (step 3). Specific T cells precursors clonalicaly expand in draining lymph nodes, recirculate via the blood and migrate into tissues including the skin. Elicitation phase) When the same hapten is applied on the skin, it penetrates the epidermis and is uptaken by epidermal cells, including skin LC and keratinocytes (step 5) which can present haptenated peptides to recirculaling T cells (step 6). Type 1 T cells produce cytokines, which activate resident skin cells allowing the production of inflammatory cytokines and chemokines. This is responsible for the recruitment of leucocytes from blood to skin leading to the development of skin lesions.

- they represent a minority of the chemicals among the thousands which are able to induce ACD in humans.
- they are endowed with potent proinflammatory properties, known as irritancy, due to the toxicity of the chemical. This toxicity provides a danger signal for the skin innate immune system, leading to: (i) production of inflammatory cytokines interleukin (IL)-1, tumor necrosis factor (TNF) and chemokines by skin cells, (ii) activation of skin DC which can initiate their maturation process and emigrate to draining lymph nodes.
- they can induce a primary ACD, i.e., an hapten-specific immune reaction, following a single skin contact, which has the same pathophysiology as the classical CHS reaction obtained with two hapten skin paintings.[9] That CHS occurs after a single exposure to haptens could be explained by persistence of haptens in the skin for several days after painting, allowing the recruitment of specific T cells at the site of skin sensitization.

Thus, strong haptens, through their toxicity, represent a danger signal able to potently activate innate immunity which in turn allows the development of a robust and rapid hapten-specific immunity. Alternatively, the most frequently encountered haptens, classified as moderate, weak or very weak, are much less irritant than strong haptens and may not have the same ability to activate innate immune cells. Consequently, the resulting ACD may operate through slightly distinct mechanisms.

$CD8^+$ T Cells Are Effector Cells While $CD4^+$ T Cells Behave As Regulatory Cells

The respective contribution of $CD4^+$ and $CD8^+$ T cells in CHS has been examined using different strategies: (i) in vivo depletion of normal mice with anti-CD4 and anti-CD8 mAbs; (ii) transfer of $CD4^+$ or $CD8^+$ T cells from sensitized mice into Rag°/° mice; (iii) use of MHC class I°/° ($CD8^+$ T cell-deficient) or MHC class II°/° ($CD4^+$ T cell-deficient) mice; (iv) transfer of the ability to mediate a CHS reaction by injection of purified primed $CD4^+$ and/or $CD8^+$ T cells in naïve recipients; (v) transfer of haptenated DC from MHC class I°/° or MHC class II°/° mice to induce CHS in naïve recipients. Mice genetically deficient in the CD4 or CD8 molecule (CD4°/° or CD8°/°) do not represent models of $CD4^+$ or $CD8^+$ T cell deficiency and results obtained with these mice will be discussed later (see "IV-2 CD4-deficient mice").

Adoptive transfert experiments have first highlighted that DTH to protein was transferable into MHC class II-matched recipients whereas transfer of CHS required class I-matched recipients.[10] Using in vivo depletion of $CD4^+$ and $CD8^+$ T cell subsets, Gocinski et al were the first to suggest that $CD8^+$ T cells could mediate the CHS response to DNFB and other strong haptens.[11] They further showed that $CD4^+$ T cells were endowed with down-regulatory activity, since the CHS reaction was enhanced following in vivo depletion in $CD4^+$ T cells.

Bour et al used another approach to study the contribution of $CD4^+$ and $CD8^+$ T cell subsets. They studied CHS in MHC class I and MHC class II KO mice which are deficient in $CD8^+$ and $CD4^+$ T cells, respectively.[12-14] Indeed, $CD4^+$ T cells develop during the ontogeny by interaction with thymic APCs expressing MHC class II molecules. In the absence of such MHC class II molecules the $CD4^+$ T cell pool cannot differentiate and $CD8^+$ T cells compose most of the circulating mature T cells. Likewise, due to lack of positive selection of $CD8^+$ T cells in MHC class I-deficient mice, $CD4^+$ T cells constitute the vast majority of mature peripheral T cells. Application of a CHS reaction to these mice showed surprising results. Class I °/° mice did not develop any CHS response to DNFB, indicating that $CD8^+$ T cells were mandatory for the development of the pathology. Since these I°/° mice have normal numbers and functions of $CD4^+$ T cells and can mount a classical DTH reaction to alloantigens,[12-14] these data demonstrated that $CD4^+$ T cells do not mediate the CHS reaction to DNFB. On the other hand, class II°/° mice developed an enhanced CHS reaction, with chronic skin inflammation. Moreover, in vivo depletion of $CD8^+$ T cells in MHC class II°/° mice resulted in a complete abrogation of the CHS. More importantly, development of hapten-specific $CD8^+$ effectors could occur in the absence of CD4 help. Indeed, there is no need of $CD4^+$ T cells for the priming of $CD8^+$ T cells in II°/° mice and the presence of $CD4^+$ T cells has a negative effect on the intensity of the $CD8^+$ T cell-mediated CHS response.[12,15] Thus, another important information from these studies was the characterization of MHC class II-restricted $CD4^+$ T cells as down-regulatory cells of CHS.

Most of the data summarized above have been obtained with DNFB in C57BL/6 (H2b)[12,16] and in BALB/C (H2d)[17] mice. Similar results have been obtained for other haptens such as OX,[17,18] dimethylbenzanthracene (DMBA),[19] and TNCB.[15,20] Thus, these findings indicated that a functional dichotomy exists between $CD8^+$ T cells and $CD4^+$ T cells which behave as effector cells and regulatory cells, respectively, in CHS to strong haptens. No data are available yet for moderate to weak haptens, mainly because of the lack of experimental murine model for these haptens.

Priming of Specific CD8⁺ and CD4⁺ in Lymphoid Organs During the Sensitization Phase of CHS

CD8⁺ Type 1 Cells and CD4⁺ Type 2 Cells

Priming of naïve T cell precursors occurs in the draining lymph nodes in a few days following hapten skin painting. The optimal time between hapten painting and T cell priming is 5 days in murine models. At that time T cells recovered from lymph nodes are endowed with potent proliferative activities.[16,21] Analysis of cytokine production by CD4⁺ and CD8⁺ T cell subsets after in vitro restimulation by haptenated APCs have shown that CD8⁺ T cells produce type 1 cytokines, mostly IFN-γ, while CD4⁺ T cells produce type 2 cytokines, including IL-4, IL-5 and IL-10.[18] These results were subsequently confirmed using ELISPOT assays to demonstrate high numbers of IFN-γ-producing CD8⁺ T cells[22] and IL-4- producing CD4⁺ T cells in the lymph nodes of hapten-sensitized mice.[18] Analysis of the number of DNFB-specific CD8⁺ T cells using an IFN-γ ELISPOT assay showed an average of 50 CD8⁺ T cell precursors/ 10^5 lymph node cells at day 5 post sensitization, a number which is similar to what is found in other antigen-specific immune responses.[22]

MHC Restriction of Hapten-Specific T Cells

Investigators from different groups provided evidence that hapten presentation to T cell in CHS was MHC restricted and thus similar to the presentation of protein antigen-derived peptides in classical DTH.[15,20] Immunization of mice with hapten-pulsed DC recovered from the epidermis or derived from bone-marrow precursors is able to prime for specific T cells which proliferate to DNFB in secondary proliferative responses. Immunization procedures using DC recovered from MHC class I or MHC class II-deficient mice confirmed the opposite functional effects of the CD8 and CD4 T cell pools.[16] In these experiments, MHC class I-expressing DC (either from normal mice or from MHC I⁺/II⁻ mice) induced the priming of CD8⁺ T cells in the lymph nodes (assessed by specific proliferation) and the CHS reaction upon subsequent challenge. Conversely, immunization by DC lacking MHC class I molecules (recovered from MHC class I- deficient mice) was inefficient at inducing a CHS reaction but could prime for CD4⁺ T cells. Indeed, the CD4⁺ T cells purified from the lymph nodes of such mice were hapten-specific, as assessed in secondary proliferative responses.[16]

These results were confirmed by a recent study in non genetically modified mice, using bone-marrow-derived DC which were pulsed with trinitrophenyl (TNP)-derivatized peptides and administred intradermally to generate a CHS reaction. Two types of peptides that have affinity for either MHC class I or class II peptides were used. Martin et al showed that the class I binding peptides induced CHS responses similar to that obtained with epicutaneous TNP application. In contrast, DC pulsed with class II binding peptides did not sensitize for optimal CHS.[23] On this basis, Cavani et al speculated that the ability of chemical haptens to drive CD8⁺ T cells activation may be associated with their capacity to interact directly with peptides bound within MHC-I or MHC-II to create immunogenic trimolecular complexes.[5]

CD8⁺ T Cell Priming Does Not Require CD4⁺ T Cell Help

Classically, optimal activation of naïve CD8⁺ T cells requires signals received by CD4⁺ T cells, and referred to as CD4⁺ T cell help. Two models have been proposed which provide a general framework for the role of CD4⁺ T cells in mediating help for CTLs. In the "three-cell" model, help is provided to CTLs by CD4⁺ T cells that recognize Ag on the same APC. In the sequential "two-cell" model CD4⁺ T cells first interact with APC, which in turn activate naive CTLs. This CD4⁺ T cell help involves the CD40-CD40L interaction. Indeed, more than IL-2 synthesis, expression of CD40-L by activated $CD4^+$ T cells may be one necessary and sufficient factor to fulfill the helper function.[24]

In CHS to strong haptens, $CD8^+$ T cell activation in the lymph nodes does not require $CD4^+$ T cell help and involvement of the CD40/CD40-L is unlikely, since CD40-L-deficient mice mount a normal CHS to DNFB.[25] Indeed, immunization of mice which are deficient in $CD4^+$ T cells either by in vivo treatment with monoclonal antibodies or because they lack the MHC class II-restricted $CD4^+$ T cell compartment (MHC class II KO mice) develop a strong CHS response to haptens. Similarly, the immunization studies performed with either MHC I^+/II° dendritic cells[16] or with I^+/II^+ DC haptenized with MHC class I binding peptides[23] have confirmed that CHS could develop without the activation of MHC class II-restricted $CD4^+$ T cell pool.

Other studies on viral-induced DTH responses have indicated that activation of naive $CD8^+$ T cells for the generation of MHC class I-restricted immune responses can occur in the absence of T cell help.[26,27] Recently, it was demonstrated that the main parameter which dictates the requirement or the absence of requirement of CD4 help was the number of CTL precursors which could be activated at time of priming[28,29] Indeed, CTL responses induced by cross-priming can be converted from CD4-dependent to CD4-independent by increasing the frequency of CTL precursors. In the absence of CD4 T cells, high numbers of CTL precursors were able to expand and become effector CTLs.[29] The ability of high frequencies of CD8 T cells to override help was not due to their ability to signal CD40 via expression of CD154. These findings suggest that when precursor frequencies are high, priming of CD8 T cell responses may not require CD4 T cell help.

Another explanation for the development of $CD8^+$ effector cells is that antigens which have intrinsic ability to induce DC maturation bypass the need for CD4 help via CD40 activation.[30] Indeed, mice depleted in $CD4^+$ T cells can be primed for CTL responses by transfer of LPS-activated, antigen-pulsed DC.

In CHS, DC maturation induced by haptens with strong inflammatory capacities may bypass the need for CD4 help via CD40/CD40-L interaction and may be sufficient to trigger specific CTL responses with a high precursor frequency.

Elicitation Phase of CHS Is Due to the Recruitment and Activation of $CD8^+$ CTLs

Since the main function of $CD8^+$ T cells is cytotoxicity, the observation that CHS was mediated by $CD8^+$ cells raised the possibility that cytotoxicity was mandatory for expression of CHS. $CD8^+$ CTLs are effector cells of the immune defence system again viruses and tumors[31] and exert their lytic functions through 2 main independant mechanisms.[32] The secretory pathway involves the release of perforin and granzymes from cytolytic granules. The nonsecretory pathway involves interaction of the Fas ligand (FasL) upregulated during T cell activation, with the apoptosis-inducing Fas molecule on the target cell.

Preliminary experiments using mice deficient in either perforin or FasL were disappointing since perforin-KO mice and FasL (gld) – deficient mice developed a normal CHS reaction to DNFB and contained $CD8^+$ CTLs able to kill haptenated targets. However, FasL- and perforine-double deficient mice could not develop CHS suggesting that cytotoxicity was necessary for the development of the pathologic process and that one cytotoxic pathway could compensate the absence of the other one.[22] This hypothesis was confirmed by extensive in vitro studies which demonstrated that primed $CD8^+$ T cells from perforine- or FasL-KO mice could kill haptenized targets whereas no cytotoxicity was observed using $CD8^+$ T cells from double deficient mice. Finally, the observation that perforine-KO CTLs could not kill Fas-deficient targets cells confirmed that both cytotoxic pathways were involved in the anti-hapten CTL activity.[22]

The involvement of cytotoxic CTLs in the development of CHS was analyzed during the elicitation phase by following the migration of $CD8^+$ T cells. Akiba et al could demonstrate that $CD8^+$ T cells could infiltrate the challenged skin as early as 9 hours after skin painting and that this migration was associated with IFN-γ production and induction of apoptosis in skin epidermal cells.[17] Double staining experiments using MHC class II antibodies and TUNEL staining showed that keratinocytes were the main target of CTLs. Thus, $CD8^+$ T cells are endowed with in vivo cytotoxic activity and keratinocytes behave as antigen-presenting cells during the elicitation phase of CHS. The contribution of other cell types in the activation of hapten-specific CTLs remain however unknown. Since MHC class I molecules are expressed on all cells, it is highly probable that haptens which rapidely diffuse through the epidermis could be expressed as haptenated peptides by different skin cell types. In this respect Biderman et al reported that mastocyte activation and chemokine production was needed for the recruitment of neutrophils which are necessary for development of CHS.[33] Since neutrophils enter the skin after $CD8^+$ T cells, it is possible that mast cell activation could be secondary to CTL activation by presentation of haptenated peptides by mast cells.

Apoptosis is involved in several skin pathologies and is not restricted to CHS. In ACD, several reports have emphasized the existence of apoptotic processes involving the epidermis.[4] More recently, Akdis et al have demonstrated that skin lesions of atopic dermatitis were associated to the occurrence of massive apoptosis of epidermal cells.[34] Although the contribution of CTLs in the pathophysiology of atopic dermatitis (AD) is not known precisely, these data emphasize that epidermal cell apoptosis is a common feature of eczematous dermatoses and suggest that anti-apoptic drugs could be new therapeutic tools.[35]

$CD4^+$ T Cells Down-Regulate the CHS Reaction

Regulatory cells are key actors in maintaining peripheral tolerance and controlling inflammatory responses. They are endowed with the capacity to inhibit the development of a potentially dangerous immune response have been described in many models. The current complexity is in part due to the diversity of models used, which have enabled the identification of a regulatory component in almost every T cell subset and which have brought into evidence many much-debated regulatory mechanisms.[5,36,37] Three main regulatory $CD4^+$ T cell subsets have been studied: (i) Tr1 cell clones which produce high amounts of the immunosuppressive cytokine IL-10; (ii) Type 2 $CD4^+$ T cells which polarize T cells towards a type 2 phenotype and antagonize the type 1 biais characteristic of CHS; (iii) and naturally occurring $CD4^+CD25^+$ T cells.

From the studies described above, $CD4^+$ T lymphocytes behave as down-regulatory cells and most likely regulate both the sensitization and elicitation phases of CHS (Fig. 2).

Within secondary lymphoid organs, following sensitization, $CD4^+$ cells limit the size of the $CD8^+$ effector cell pool[14] or modify their functional properties. The number of specific $CD8^+$ T cells, determined by IFN-γ ELISOPT assay, is much higher in $CD4^+$ T cell deficient mice than in normal mice, suggesting that $CD4^+$ T cells control the development of the $CD8^+$ T cell pool.[38]

After migrating to the challenge site, these cells probably contribute to the control of inflammation and its resolution.[17] Indeed, in the absence of $CD4^+$ T cells, mice develop a more pronounced and persistent inflammation.[11,12,18,39] It is noteworthy that $CD4^+$ T cells are recruited in challenged skin hours after recruitment of $CD8^+$ T cells,[17] suggesting that their entry into the skin is responsible for the inactivation of $CD8^+$ CTL activity, thereby limitating the development of the inflammatory pathogenic process.

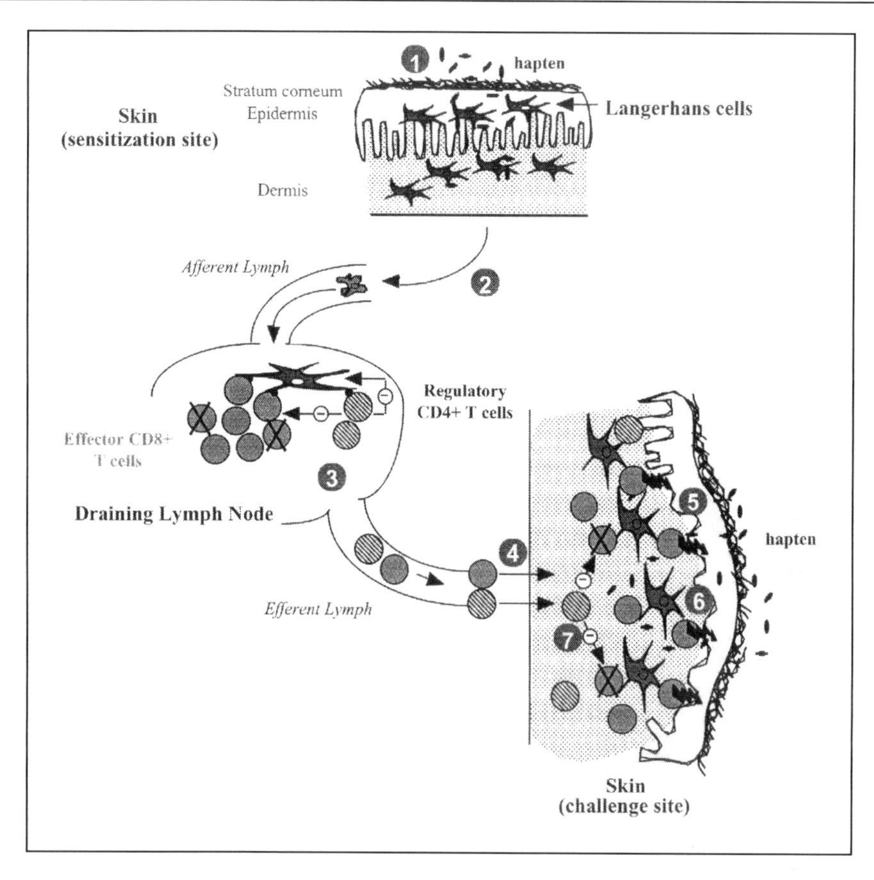

Figure 2. CHS to strong haptens. Sensitization phase) Haptens penetrate the epidermis (step 1) and are uptaken by skin DC which migrate to the draining lymph nodes (step 2) where they present haptenated peptides to both $CD8^+$ effectors and down-regulatory $CD4^+$ T cells (step 3). Specific T cells precursors clonalicaly expand in draining lymph nodes, recirculate via the blood and migrate to tissues including the skin. Elicitation phase) When the same hapten is applied on the skin, it penetrates the epidermis and is uptaken by epidermal cells, including skin LC and keratinocytes (step 5). Activation of $CD8^+$ CTLs induce apoptosis of keratinocytes and production of cytokines and chemokines by skin resident cells (step 6). This induces the recruitment of leucocytes from the blood to the skin. $CD4^+$ T cells may block the $CD8^+$ T cell activation, thereby inducing the down-regulation of CHS (step 7).

Limited information is currently available regarding whether a particular subset of regulatory cells is involved in the regulation of CHS. Nickel specific Tr1 cells (producing high amounts of IL-10) have been cloned from skin lesions of ACD patients suggesting that this subset of regulatory cells might contribute to the regulation of the efferent phase of contact sensitivity.[8] Indirect evidence for the implication of $CD4^+CD25^+$ cells comes from the observation that IL-2-IgG2b fusion protein inhibited CHS associated with an increase of the size of the $CD4^+CD25^+$ T cell compartment.[40] Our own data, in the model of CHS to DNFB support a role for $CD4^+CD25^+$ regulatory T cells in the control of the response and in the establishment of oral tolerance (B. Dubois and D. Kaiserlian, submitted).

CD4$^+$ T Cells May Be Effector Cells of CHS in Some Experimental Conditions

Although most of recent studies have emphasized the major effector role of CD8$^+$ T cells in CHS, it cannot be concluded that CD4$^+$ T cells and other cells are unable to act as CHS effectors.

Particularity of Some Chemicals

As discussed above, there are no data on the effector cells of CHS to moderate, weak and very weak haptens. The main reason is because there are no reproducible animal model for these weak haptens. However, since weak haptens have very limited irritant properties and thus probably do not activate the innate immunity as do strong haptens, it is possible that the CHS response they induce could be mediated by both CD4$^+$ and CD8$^+$ T cells. It is also tempting to speculate that the number of specific T cell precursors will be lower and that CD4 T cell help would be required for optimal CHS responses.

Some chemicals, e.g., FITC and formaldehyde, seem to provoke preferential type 2 cytokine production by CD4$^+$ T cells, which have been shown to mediate the CHS reaction.[41] However, experiments using either depletion of T cell subets or mice deficient in CD4$^+$ or CD8$^+$ T cells are still lacking to sustain this hypothesis.

CD4-Deficient Mice

That CD4$^+$ T cells were effectors of CHS was concluded from studies of Kondo et al and Wang et al showing that CHS to DNFB and OX was greatly impaired in CD4°/° mice, genetically deficient in the CD4 molecule.[42,43] The reason for the discrepancy between these results and those reported by other investigators most likely reflects important functional differences between CD4-deficient and MHC class II-deficient mice. Indeed, although the CD4 gene has been knocked, cells exerting a helper cell activity can be recovered from the double negative, CD4$^-$8$^-$ T cell subset, in CD4°/° mice. In this respect, it is noteworthy that efficient thymic maturation of helper T cells has been shown in these CD4°/° mice.[44] The CD4$^-$ helper T cells express αβ-TCR and are able to control Leishmania infections, to mediate antibody class switch and DTH reaction to keyhole limpet hemocyanin (KLH),[44-46] Thus, although CD4°/° mice do not have the CD4 molecule, they are able to mount MHC class II restricted reactions, suggesting that the CD4 molecule is not absolutely required for efficient recognition of antigens presented by MHC class II molecules on antigen presenting cells. To understand why CD4°/° mice cannot develop a normal CHS response to DNFB, we set up a series of experiments concerning hapten-specific CTL activity in this particular mice (P. Saint-Mezard, manuscript submitted). We show that the absence of the CD4 molecule does not impair the priming of hapten-specific IFN-γ-producing CD8$^+$ T cells but dramatically reduce the development of cytotoxic activity. More importantly, deficient specific CTL activity could be restored by restimulating CD8$^+$ T cells by class II-deficient APCs, suggesting that the development of specific cytotoxic activity in CD4°/° mice is blocked by an MHC class II restricted population. We hypothesize that the particular MHC class II-restricted population present in CD4°/° mice may assume a regulatory function toward the CHS reaction, as does the CD4$^+$ T cell population in normal mice, and that this role may be exacerbated in this particular KO mouse. Collectively, the data from the CD4°/° mice support the concept that CD8$^+$ T cells are effector cells of CHS and illustrate a role for the CD4 molecule in the development of a down-regulatory T cell population.

CD4⁺ T Cells May Be Effector Cells in CHS to Haptens When CD8⁺ T Cells Are Deficient

CD4⁺ T cells could be effector cells in CHS to some haptens when the CD8⁺ T cell population is deficient. Evidence came from studies of Martin et al.[23] They first showed that dendritic cells pulsed with TNP-derivatized peptides that have affinity for class II molecules could induce a low, albiet significant CHS reaction.[23] Next, they used C57BL/6 mice and class I°/° mice and studied the CHS to DNP and TNP. CHS to DNP was normal in C57BL/6 mice and absent in I°/° mice as previously reported.[12] TNP was able to induce a CHS response in C57BL/6 which was inhibited by in vivo depletion of CD8⁺ T cells using specific mAbs. Surprisingly, CD8⁺ T cell-deficient I°/° mice were able to develop a CHS reaction to TNP, which was similar to the TNP response in C57BL/6 mice in its kinetics (Martin et al, personal communication). These data show that, in the absence of CD8⁺ T cells, and for some but not all haptens, a CHS response can be mediated by CD4⁺ T cells.

Role of Other Cell Types in CHS

Although the αβ T cells are responsible for the hapten-specific CHS reaction, other lymphoid cell subsets have been shown to be implicated in the complex process of cellular activation required for optimal development of CHS.[47,48]

B-1 cells are activated in lyphoid organs during the sensitization phase and produce IgM antibodies. These antibodies diffuse in the skin and will bind the hapten immediatly after the challenge, leading to complement activation which seems mandatory for the recruitment of effector T cells at the challenge site. Natural killer (NK) T cells seem to be important for the activation of the B-1 cells through the production of a burst of IL-4. γδ T cells, known as dendritic epidermal T cells (DETC), are necessary for activation and function of specific αβ T cells once they are recruited into the challenged site.

CD8⁺ and CD4⁺ T Cells in Human ACD

Most of the hypotheses on the precise mechanisms which may lead to human allergic contact dermatitis come from studies of the murine CHS reaction. Thus, the current pathophysiology of ACD postulates that CD8⁺ T cells are the main effector cells, while the CD4⁺ T cell pool comprises down-regulatory cells, able to block effector cell functions.[5,8]

T cells involved in ACD have been recovered from the blood or the skin of ACD patients. They are extremely heterogenous in their cytokine profile and function, with CD4⁺ and CD8⁺ T cells thought to play distinct roles in the development of the inflammatory response. For example, clinical studies examining long-term T cell clones generated from the peripheral blood and skin lesions of patients with allergic contact dermatitis (ACD) have yielded support for both CD4⁺ and CD8⁺ T cells as mediators of ACD in humans. Clones generated from lesions of patients with nickel-mediated contact dermatitis were CD4⁺T cells, implicating these cells as effector cells in this pathology.[49] In more recent study, expression of ACD to nickel correlates with the frequency of specific CD8⁺ T cells in the peripheral blood, which is high in allergic individuals. In contrast, the peripheral blood of both allergic and non allergic subjects shows comparable nickel-reactive CD4⁺ T cells responses.[50] Similarly, clinical studies of patients with contact allergies to classical haptens such as urushiol have demonstratedd that most of the hapten-specific T cells isolated from the patient lesions were IFN-γ-producing CD8⁺ T cells.

Most, but not all, human hapten-specific CD8⁺ T cells display a type 1 cytokines profile, whereas CD4⁺ T cells isolated from ACD lesions show a more variable pattern of cytokine release, with a predominance of Th1 cells and a lower number of Th2 cells. Tr1 lymphocytes represent 7-10% of nickel-specific T-cell clones isolated from ACD skin or the blood of allergic individuals, and their number is higher in the blood of non allergic subjects. In a manner dependent on IL-10, these Tr1 cells block the maturation of DCs and the release of IL-12, thus impairing the capacity of DCs to activate hapten-specific Th1 effector lymphocytes.[8]

Conclusions

In summary, CHS reaction and ACD can be viewed as the result of activation of two distinct T cell subsets endowed with opposite functions: effector T cells and down-regulatory T cells. The severity and the duration of the skin inflammation appears directly related to the respective activation state and/or size of these two compartments. Thus, overwhelming regulation in sensitized individuals may lead to lack of inflammation (tolerance) despite repeated exposures to the hapten, while defect in regulatory cells may explain chronic contact dermatitis. Further studies will have to adress for the possibility of reversing an established ACD by either targetting the effector T cell population or by increasing the number or functionnal properties of regulatory T cells.

References

1. Moskophidis D, Lehmann-Grube F. Virus-induced delayed-type hypersensitivity reaction is sequentially mediated by $CD8^+$ and $CD4^+$ T lymphocytes. Proc Natl Acad Sci USA 1989; 86:3291-3295.
2. Kundig TM, Althage A, Hengartner H et al. Skin test to assess virus-specific cytotoxic T-cell activity. Proc Natl Acad Sci USA 1992; 89:7757-7761.
3. Kalish RS, Askenase PW. Molecular mechanisms of CD8+ T cell-mediated delayed hypersensitivity: Implications for allergies, asthma, and autoimmunity. J Allergy Clin Immunol 1999; 103:192-199.
4. Krasteva M, Kehren J, Ducluzeau MT et al. Contact dermatitis I. Pathophysiology of contact sensitivity. Eur J Dermatol 1999; 9:65-77.
5. Cavani A, Albanesi C, Traidl C et al. Effector and regulatory T cells in allergic contact dermatitis. Trends Immunol 2001; 22:118-120.
6. Gorbachev AV, Fairchild RL. Regulatory role of CD4+ T cells during the development of contact hypersensitivity responses. Immunol Res 2001; 24:69-77.
7. Watanabe H, Unger M, Tuvel B et al. Contact hypersensitivity: The mechanism of immune responses and T cell balance. J Interferon Cytokine Res 2002; 22:407-412.
8. Girolomoni G, Sebastiani S, Albanesi C et al. T-cell subpopulations in the development of atopic and contact allergy. Curr Opin Immunol 2001; 13:733-737.
9. Saint-Mezard P, Krasteva M, Chavagnac C et al. Afferent and Efferent Phases of Allergic Contact Dermatitis (ACD) can be induced after a single skin contact with hapten: Evidence using a mouse model of primary ACD. J Invest Dermatol 2003; 120:641-647.
10. Sunday ME, Dorf ME. Hapten-specific T cell response to 4-hydroxy-3-nitrophenyl acetyl. X. Characterization of distinct T cell subsets mediating cutaneous sensitivity responses. J Immunol 1981; 127:766-768.
11. Gocinski BL, Tigelaar RE. Roles of CD4+ and CD8+ T cells in murine contact sensitivity revealed by in vivo monoclonal antibody depletion. J Immunol 1990; 144:4121-4128.
12. Bour H, Peyron E, Gaucherand M et al. Major histocompatibility complex class I-restricted CD8+ T cells and class II-restricted CD4+ T cells, respectively, mediate and regulate contact sensitivity to dinitrofluorobenzene. Eur J Immunol 1995; 25:3006-3010.
13. Cosgrove D, Gray D, Dierich A et al. Mice lacking MHC class II molecules. Cell 1991; 66:1051-1066.
14. Chan SH, Cosgrove D, Waltzinger C et al. Another view of the selective model of thymocyte selection. Cell 1993; 73:225-236.
15. Bouloc A, Cavani A. Contact hypersensitivity in MHC class II-deficient mice depends on CD8 T lymphocytes primed by immunostimulating Langerhans cells. J Invest Dermatol 1998; 111:44-49.

16. Krasteva M, Kehren J, Horand F et al. Dual role of dendritic cells in the induction and down-regulation of antigen-specific cutaneous inflammation. J Immunol 1998; 160:1181-1190.
17. Akiba H, Kehren J, Ducluzeau MT et al. Skin inflammation during contact hypersensitivity is mediated by early recruitment of CD8+ T cytotoxic 1 cells inducing keratinocyte apoptosis. J Immunol 2002; 168:3079-3087.
18. Xu H, DiIulio NA, Fairchild RL. T cell populations primed by hapten sensitization in contact sensitivity are distinguished by polarized patterns of cytokine production: Interferon gamma-producing (Tc1) effector CD8+ T cells and interleukin (Il) 4/Il-10-producing (Th2) negative regulatory CD4+ T cells. J Exp Med 1996; 183:1001-1012.
19. Anderson C, Hehr A, Robbins R et al. Metabolic requirements for induction of contact hypersensitivity to immunotoxic polyaromatic hydrocarbons. J Immunol 1995; 155:3530-3537.
20. Kolesaric A, Stingl G, Elbe-Burger A. MHC class I+/II- dendritic cells induce hapten-specific immune responses in vitro and in vivo. J Invest Dermatol 1997; 109:580-585.
21. Fehr BS, Takashima A, Matsue H et al. Contact sensitization induces proliferation of heterogeneous populations of hapten-specific T cells. Exp Dermatol 1994; 3:189-197.
22. Kehren J, Desvignes C, Krasteva M et al. Cytotoxicity is mandatory for CD8(+) T cell-mediated contact hypersensitivity. J Exp Med 1999; 189:779-786.
23. Martin S, Lappin MB, Kohler J et al. Peptide immunization indicates that CD8+ T cells are the dominant effector cells in trinitrophenyl-specific contact hypersensitivity. J Invest Dermatol 2000; 115:260-266.
24. Ridge JP, Di Rosa F, Matzinger P. A conditioned dendritic cell can be a temporal bridge between a CD4+ T-helper and a T-killer cell. Nature 1998; 393:474-478.
25. Gorbachev AV, Heeger PS, Fairchild RL. CD4+ and CD8+ T cell priming for contact hypersensitivity occurs independently of CD40-CD154 interactions. J Immunol 2001; 166:2323-2332.
26. Buller RM, Holmes KL, Hugin A et al. Induction of cytotoxic T-cell responses in vivo in the absence of CD4 helper cells. Nature 1987; 328:77-79.
27. Rahemtulla A, Fung-Leung WP, Schilham MW et al. Normal development and function of CD8+ cells but markedly decreased helper cell activity in mice lacking CD4. Nature 1991; 353:180-184.
28. Wang B, Norbury CC, Greenwood R et al. Multiple paths for activation of naive CD8+ T cells: CD4-independent help. J Immunol 2001; 167:1283-1289.
29. Mintern JD, Davey GM, Belz GT et al. Cutting edge: precursor frequency affects the helper dependence of cytotoxic T cells. J Immunol 2002; 168:977-980.
30. Schuurhuis DH, Laban S, Toes RE et al. Immature dendritic cells acquire CD8(+) cytotoxic T lymphocyte priming capacity upon activation by T helper cell-independent or -dependent stimuli. J Exp Med 2000; 192:145-150.
31. Berke G: The CTL's kiss of death. Cell 1995; 81:9-12.
32. Corazza N, Muller S, Brunner T et al. Differential contribution of Fas- and perforin-mediated mechanisms to the cell-mediated cytotoxic activity of naive and in vivo-primed intestinal intraepithelial lymphocytes. J Immunol 2000; 164:398-403.
33. Biedermann T, Kneilling M, Mailhammer R et al. Mast cells control neutrophil recruitment during T cell-mediated delayed-type hypersensitivity reactions through tumor necrosis factor and macrophage inflammatory protein 2. J Exp Med 2000; 192:1441-1452.
34. Trautmann A, Akdis M, Kleemann D et al. T cell-mediated Fas-induced keratinocyte apoptosis plays a key pathogenetic role in eczematous dermatitis. J Clin Invest 2000; 106:25-35.
35. Trautmann A, Akdis M, Schmid-Grendelmeier P et al. Targeting keratinocyte apoptosis in the treatment of atopic dermatitis and allergic contact dermatitis. J Allergy Clin Immunol 2001; 108:839-846.
36. Dubois B, Chapat L, Goubier A et al. CD4+CD25+ T cells as key regulators of immune responses. Eur J Dermatol 2003; 13:111-116.
37. Grabbe S, Schwarz T. Immunoregulatory mechanisms involved in elicitation of allergic contact hypersensitivity. Immunol Today 1998; 19:37-44.
38. Desvignes C, Etchart N, Kehren J et al. Oral administration of hapten inhibits in vivo induction of specific cytotoxic CD8+ T cells mediating tissue inflammation: a role for regulatory CD4+ T cells. J Immunol 2000; 164:2515-2522.

39. Desvignes C, Bour H, Nicolas JF et al. Lack of oral tolerance but oral priming for contact sensitivity to dinitrofluorobenzene in major histocompatibility complex class II-deficient mice and in CD4+ T cell-depleted mice. Eur J Immunol 1996; 26:1756-1761.
40. Ruckert R, Brandt K, Hofmann U et al. IL-2-IgG2b fusion protein suppresses murine contact hypersensitivity in vivo. J Invest Dermatol 2002; 119:370-376.
41. Dearman RJ, Kimber I: Role of CD4(+) T helper 2-type cells in cutaneous inflammatory responses induced by fluorescein isothiocyanate. Immunology 2000; 101:442-451.
42. Kondo S, Beissert S, Wang B et al. Hyporesponsiveness in contact hypersensitivity and irritant contact dermatitis in CD4 gene targeted mouse. J Invest Dermatol 1996; 106:993-1000.
43. Wang B, Fujisawa H, Zhuang L et al. CD4+ Th1 and CD8+ type 1 cytotoxic T cells both play a crucial role in the full development of contact hypersensitivity. J Immunol 2000; 165:6783-6790.
44. Bachmann MF, Oxenius A, Mak TW et al. T cell development in CD8-/- mice. Thymic positive selection is biased toward the helper phenotype. J Immunol 1995; 155:3727-3733.
45. Hornquist CE, Ekman L, Grdic KD et al. Paradoxical IgA immunity in CD4-deficient mice. Lack of cholera toxin-specific protective immunity despite normal gut mucosal IgA differentiation. J Immunol 1995; 155:2877-2887.
46. Locksley RM, Reiner SL, Hatam F et al. Helper T cells without CD4: Control of leishmaniasis in CD4-deficient mice. Science 1993; 261:1448-1451.
47. Askenase PW. Yes T cells, but three different T cells (alphabeta, gammadelta and NK T cells), and also B-1 cells mediate contact sensitivity. Clin Exp Immunol 2001; 125:345-350.
48. Yokozeki H, Watanabe K, Igawa K et al. Gammadelta T cells assist alphabeta T cells in the adoptive transfer of contact hypersensitivity to para-phenylenediamine. Clin Exp Immunol 2001; 125:351-359.
49. Sinigaglia F, Scheidegger D, Garotta G et al. Isolation and characterization of Ni-specific T cell clones from patients with Ni-contact dermatitis. J Immunol 1985; 135:3929-3932.
50. Cavani A, Mei D, Guerra E et al. Patients with allergic contact dermatitis to nickel and nonallergic individuals display different nickel-specific T cell responses. Evidence for the presence of effector CD8+ and regulatory CD4+ T cells. J Invest Dermatol 1998; 111:621-628.

Chapter 5

T Cell Subsets in Allergic Contact Dermatitis

Andrea Cavani, Francesca Nasorri and Giampiero Girolomoni

Summary

Allergic contact dermatitis is the result of an exaggerated immune response sustained by $CD8^+$ and $CD4^+$ type 1 T lymphocytes towards small chemicals penetrating through the skin. Expression of the disease depends upon a coordinated series of events, which include leukocyte extravasation and positioning at the site of hapten penetration, activation of hapten-specific T cells, T cell-mediated apoptosis of keratinocytes and release of pro inflammatory cytokines, which modulate the amplification and the persistence of the inflammatory response. Termination of allergic contact dermatitis appears as a necessary event to avoid excessive tissue damage, and is the consequence of both exhaustion of effector T cells and intervention of specialized T cell subsets with regulatory functions. Among these, interleukin (IL)-10 producing T regulatory cells 1 and $CD4^+CD25^+$ T lymphocytes have been shown to be actively recruited in the skin after hapten challenge and to regulate the magnitude of the allergic reaction.

Introduction

Allergic contact dermatitis (ACD) occurs in sensitized persons as a T cell-mediated immune response to chemically reactive haptens applied to the skin surface.[1-3] T lymphocytes involved in ACD belong to a very special subset of memory-effector T cells with the competence to circulate in the cutaneous environment. This commitment is defined by the repertoire of homing and chemokine receptors acquired during the differentiation process and maintained during the T cell life-span.[4-8] Perturbing stimuli applied onto the skin such as chemicals, irritants, bacterial and viral products activate resident cells, promoting the synthesis and expression of selectins, adhesion molecules and membrane-bound chemokines on endothelial cells. Changes in the skin microvasculature, together with the local release of chemotactic substances are responsible for the augmented adhesiveness and extravasation of leukocytes in the peripheral tissue.[9-12] Thus, the rapid T cell recruitment at the challenged site appears as the consequence of nonspecific pro-inflammatory mechanisms that act to preserve skin homeostasis. Full expression of ACD, however, requires the specific activation of hapten-reactive T cells, responsible for cytolytic events and cytokine production. Amplification of the cutaneous inflammation mostly depends upon the cross-talk between T lymphocytes and resident cells, which are fully activated in response to T cell-derived cytokines and direct further massive leukocyte extravasation at the site of hapten challenge.[13-16] The current held view is that multiple mechanisms tightly regulate ACD expression. In particular, specialized subsets of $CD4^+$ lymphocytes with suppressive function have been isolated and characterized both in mice and

Immune Mechanisms in Allergic Contact Dermatitis, edited by Andrea Cavani and Giampiero Girolomoni. ©2005 Eurekah.com.

humans.[17-19] These cells may either prevent undesired immune responses to harmful antigens applied onto the skin, or promote the termination of ongoing immune responses by dampening T cell activation at the site of hapten application. Thus distinct T cell subsets, which differ profoundly in their functions, orchestrate immune responses to haptens and determine whether silent immune responses or pathologic and excessive responses to skin applied chemicals occurs in each individual at a given time.

T Cell Populations in ACD

T cells infiltrating acute ACD, as well as hapten-specific T cells isolated from peripheral blood of sensitized individuals, vary noticeably in terms of phenotype, cytokine profile, MHC-restriction and antigen-presenting cell (APC) requirement for efficient activation. Heterogeneity of the T cell infiltrate mostly refers to the $CD4^+$ population, which comprise effector T helper (Th) 1 cells, Th0 and Th2 cells, and other subsets with prominent regulatory function, such as the T regulatory (Tr)1 cells and the $CD4^+CD25^+$ lymphocytes.[20-24] In contrast, most of the $CD8^+$ lymphocytes involved in ACD express a type 1 cytokine profile, although hapten-specific T cytotoxic (Tc) 2 cells and $CD8^+$ lymphocytes secreting elevated amount of IL-10 have also been isolated from allergic individuals (personal observation). In general, a more clear-cut polarization of hapten-specific T cells can be observed in murine contact hypersensitivity (CHS), compared to human ACD. This finding, rather than reflecting differences related to the species, may be the consequence of the sensitization procedure, which is experimentally induced with a single dose of strong sensitizers in mouse, and is frequently the outcome of repeated contact with subliminal concentration of weak sensitizers in human. The profile of hapten-specific T cells may also vary in time in each single individual, it may depend on the site of isolation of the T cell population, and is influenced by the hapten to which the response is directed. Concerning the pattern of cytokine release, differences between blood- and skin-derived hapten-specific T cells have been described, with the latter releasing higher amount of IL-4.[25] This observation either suggest that cytokine secretion of infiltrating T lymphocytes can be influenced by the skin micro-environment, or that rapid depletion of effector Th1 cells occurs at the site of hapten challenge, as a consequence of activation induced cell death. Finally, Th2 cells have been shown to dominate murine CHS to FITC, and to be increased in number in mice chronically stimulated with strong sensitizers.[26,27] The latter finding suggests a Th1 to Th2 switch in chronic immune responses to haptens. Indeed, in human ACD to nickel, which results from chronic and repeated stimulation with the metal in the everyday life, Th2 cells only accounts for 10-15%, whereas Th1 cells range from 45 to 60% of the total number of T cell clones isolated from peripheral blood or ACD skin of nickel-allergic persons.[22]

Hapten-specific T cells are also polymorphic in terms of the hapten determinant recognized by their T cell receptor (TCR), and APC requirement for efficient activation.[28-32] Such a diversity may correspond to the property of haptens to generate both processing-dependent or independent hapten-epitopes, as a consequence of their capacity to bind numerous skin and cell-surface proteins, as well as to form conjugates with peptides mounted on MHC molecules exposed on the surface of skin APCs.[33,34] For example, $CD4^+$ T lymphocytes involved in ACD to urushiol (poison ivy) recognize one or more epitopes arising from metabolic intervention of APCs, whereas urushiol-specific $CD8^+$ cells comprise both processing-dependent and -independent T lymphocytes.[35] In contrast, nickel recognition by most, but not all, $CD4^+$ and $CD8^+$ T cells is independent from APC processing.[34,36] In spite of the diversity of hapten-epitopes recognition by specific T cells, preferential usage of TCR Vβ17 in nickel-specific $CD4^+$ T cells has been reported by more than one laboratory, and their relative frequency has been correlated with disease severity.[37,38] The requirement of a metabolic intervention for the generation of some hapten-epitopes also explains the diverse capacity of distinct APC populations to activate subsets of hapten-specific T cells, with B cells being much less efficient or completely incapable to activate selective nickel-specific T cell clones compared to skin-derived dendritic cells (DC).[39] Although DC are required for efficient priming of hapten-specific naive T cells, their role in the activation of memory-effector lymphocytes during the efferent phase of ACD is less defined. It is known that the need for professional APC in the activation of effector T lymphocytes is less stringent compared to that of naive T cells, and that the effector phase of CHS is less dependent on APC-driven costimulation compared to the sensitization phase.[40,41] Indeed, DC removal by steroid applications increases the magnitude of murine CHS, indicating that other cell types may contribute to the activation of emigrated T cells.[42] Endothelial cells, fibroblasts and keratinocytes express MHC class II and up regulate MHC class I molecules when exposed to interferon (IFN)-γ and/or tumor necrosis factor (TNF)-α, and may thus function as APCs for memory $CD4^+$ and $CD8^+$ T cells.[43,44] However, it has been shown that keratinocyte and endothelial cell presentation of hapten determinants lead to a non productive T cell activation or induce specific T cell unresponsiveness.[43-45] Recently activated human, but not murine, T lymphocytes express MHC class II and functional B7 molecules and may thus potentially act as APCs for memory/effector $CD4^+$ T cells engaged in peripheral immune responses.[46-48] Indeed, in the case of nickel-specific T cell responses, about one third of specific T cell clones isolated from patients with nickel-allergy do not require the presence of professional APC for productive activation.[36] This APC-independent T cell subset is activated through a T-to-T cell hapten-presentation which is independent form epitope processing and, although less efficient than DC-T cell activation, does not ensue into T cell anergy. In this scenario, once pioneer T cells immigrated in the skin became activated by the relevant hapten, they can act simultaneously as responder T cells and APC population for other hapten-specific T lymphocytes arriving in the tissue, thus leading to a rapid amplification of the inflammatory reaction even in the absence of DCs.

Effector T Cells in ACD: Mechanisms of Tissue Damage

Changes in ACD skin predominantly consist of dermal edema, epidermal spongiosis and scattered keratinocyte shrinkage and apoptosis.[49] Although several cell types have a role in the induction of the tissue damage, T lymphocyte infiltration and subsequent activation are essential events in the expression of ACD. Early ultrastructural studies revealed damaged keratinocytes in close contact to mononuclear cells, suggesting that T cell mediated cytotoxicity could have a prominent role in ACD.[49] The critical involvement of cytolytic events in cutaneous responses to haptens have been then clearly demonstrated in $P^{0/0}$ gld mice which lacking Fas ligand (FasL) and perforin genes, both involved in T cell-mediated cytotoxicity, fail to mount CHS reactions.[50] In humans, keratinocyte apoptosis is easily detected in acute ACD by TUNEL staining, which reveals fragmented DNA in tissue sections. In sensitized persons, mRNA for perforin and FasL are strongly expressed in ACD skin as early as 12 hrs after hapten application.[51,52] It has been reported that T cell-mediated keratinocyte apoptosis mediated through the Fas-FasL mechanism is a critical event in eczematous skin reactions, and that it occurs only after prior keratinocyte exposure to IFN-γ, which induces expression of MHC class II molecules and up-regulates Fas.[53-55] Early changes observed in apoptotic keratinocytes include the cleavage of E-cadherins, adhesion molecules critically involved in keratinocyte homotypic adhesion, thus leading to the epidermal spongiosis.[56] Our studies showed that keratinocytes are highly susceptible to nickel-specific cytotoxicity induced by Tc1 and Tc2 cells, and to a lesser extent by Th1 cells, but are resistant to that of Th2 clones.[52] Interestingly, while $CD8^+$ Tc1 and Tc2 cells exert their CTL activity on both resting and IFN-γ-activated keratinocytes, Th1 cells kill exclusively keratinocytes previously exposed to IFN-γ, since they recognize MHC class

II-restricted hapten epitopes, and kill keratinocytes prevalently through the Fas-FasL pathway. In contrast, Tc1 and Tc2 lymphocytes exert their cytotoxic function mostly through the release of perforin and granzyme, and are minimal influenced by IFN-γ pretreatment of the target cells, in line with the finding that CHS responses in IFN-γ receptor-deficient mice are only partially affected.[57] Thus, the disparate cytotoxic activity of different T cell subsets against keratinocytes suggests distinct roles in the effector phase of ACD, with CD8$^+$ lymphocytes required for the initiation of the immune reaction thanks to their capacity to target resting keratinocytes, and Th1 cells exerting cytotoxic functions only at later time points, when IFN-γ released by type 1 T cells renders keratinocytes sensitive to the CD4$^+$ T cell attack. Taken together, these findings support and extent to the human disease the current view that indicated the CD8$^+$ T cell subset as the major effector population in CHS reactions, as previously demonstrated in murine models.[58-62] In line with this hypothesis, it has been shown that the frequency of peripheral blood nickel-specific CD8$^+$ correlates with the expression of disease in nickel-allergic donors.[22] In contrast, both nickel-allergic and non allergic individuals possess circulating memory CD4$^+$ cells to nickel, suggesting that either CD4$^+$ cells alone are not sufficient for the expression of the human disease or that CD4$^+$ lymphocytes from allergic and non allergic donors differ in their effector functions.[22,24,63] The role of CD4$^+$ lymphocytes in hapten-specific immune responses appears definitively more variegate compared to that of CD8$^+$ cells, and mostly related to the type of CD4$^+$ T cell considered. The contribution of CD4$^+$ cells in the expression of the disease is sustained by the finding that MHC class I deficient mice, which lack effector CD8$^+$ cells, still mount CHS reaction to haptens, although much reduced in magnitude compared to littermate controls, and transfer of murine hapten-specific Th1 cells is sufficient to induce CHS in recipients.[59-61,64] In addition, although not as efficient as Tc1 and Tc2 cells in inducing keratinocyte apoptosis, activated CD4$^+$ Th1 lymphocytes are a major source of pro-inflammatory cytokines in ACD skin, and responsible for the engagement of resident cells in the inflammatory process. Th1-derived IFN-γ and TNF-α promote keratinocyte expression of MHC and adhesion molecules, such as intercellular adhesion molecule (ICAM)-1, and secretion of chemokines, IP-10 and monocyte chemoattractant protein (MCP)-1 in particular, which direct leukocyte recruitment into the skin, leading to the full expression of the inflammatory reaction.[13-16] The role of Th2 cells in murine CHS and human ACD is still debated. Compared to Th1 lymphocytes, Th2 cells are incapable to determine keratinocytes damage, and are much less efficient in inducing activation of resident cells. Nevertheless, IL-4 synergizes with IFN-γ in promoting keratinocyte secretion of interferon-induced protein (IP)-10, and IL-17, secreted by both Th1 and Th2 cells, augment ICAM-1 expression and modulate chemokine secretion by keratinocytes.[13,16] In mouse, Th2 cells have been repeatedly reported as critical regulatory cells in CHS reactions. In particular, passive transfer of hapten-specific Th2 lymphocytes can reverse established murine CHS in a IL-4 dependent mechanism.[65] Other reports indicated that IL-4 knock out mice have a reduced CHS reaction, thus suggesting a pro-inflammatory role of IL-4.[66,67] Furthermore, CHS expression is decreased in signal transducer and activator of transcription (STAT)-6 deficient mice, in which Th2 responses are impaired.[68]

Regulatory Mechanisms in ACD

Much evidence indicates that ACD is a tightly regulated reaction. In line with this view is the fact that most individuals present memory CD4$^+$ T cells directed against environmental antigens and potential allergens without developing clinical manifestation of allergy.[20,22,63] Secondly, in allergic persons, the disease severity typically fluctuate in time,[69] and, despite the persistence of the hapten molecules in the skin, the inflammatory reaction which follows each hapten application is self-limited. Furthermore, induction of active tolerance to specific haptens occurs upon administration of the hapten through ultraviolet (UV)-irradiated mouse skin or intragastric feeding (oral tolerance).

In the last few years, numerous T cell subsets with regulatory/suppressive function have been identified.[17-19] In particular, CD4$^+$ lymphocytes which release high amount of IL-10 and IL-5 have been characterized and named Tr1 cells. Tr1 cells poorly respond to antigenic stimulation and inhibit effector CD4$^+$ and CD8$^+$ cell responses by blocking the APC functions of monocytes and DCs in a IL-10 dependent manner. Nickel-specific Tr1 cells have been isolated form the site of hapten challenge in nickel-allergic individuals, as well as from peripheral blood of both allergic and non allergic donors.[70] IL-10 released by Tr1 cells during ongoing immune responses inhibits the DC differentiation and maturation, and IL-12 release, thus impairing DC capacity to stimulate hapten-specific CD4$^+$ and CD8$^+$ T cells. In mice, evidence has been provided evidence that IL-10 is a critical cytokine for suppression or prevention of CHS reactions, and the emergence of IL-10 producing CD4$^+$ T cells have been described as responsible for hapten tolerance induced by ultraviolet radiation and in oral tolerance, although in the latter, CD8$^+$ T cells appears to play a role.[71-74]

More recently, murine and human CD4$^+$ suppressive T lymphocytes which constitutively express the CD25 antigen have been identified.[18,19] These cells have been isolated as a distinct thymic T cell lineage which is anergic upon TCR triggering in vitro, strongly express the CTLA-4 receptor, and are involved in the maintenance of peripheral tolerance to self-antigens by inhibiting auto-reactive T cells. Interestingly, about 30% of peripheral blood CD4$^+$ CD25$^+$ cells express the CLA and may be actively recruited into the skin during inflammatory processes. Indeed, CD4$^+$CD25$^+$ T cells obtained from the peripheral blood of individuals in which nickel-allergy have been excluded on the basis of a negative skin test and on clinical history of metal allergy, strongly inhibited nickel-specific CD4$^+$CD25$^-$ T cell responses.[75] These cells may thus be important in the control of undesired T cell-mediated reactions to the metal in non allergic subjects.

Whether the different suppressive T cell subsets so far identified represent distinct T cell lineages or the result of a further differentiation step from previously polarized T cells is still unknown. In the case of Tr1 cells, the latter hypothesis is suggested by the expression of markers associated with a Th1 phenotype (β2 chain of the IL-12 receptor, CD26, LAG-3),[22] and by the observation that the Tr1 phenotype is not stable, but easily revert to a Th1 profile upon repeated in vitro stimulation (personal unpublished observation). Interestingly, Tr1 cells can be induced in vitro in the presence of high levels of IL-10, or by stimulating naïve T lymphocytes with immature DCs or with DCs treated with corticosteroid and/or with vitamin D3.[76-79] Altogether, these findings strongly suggest that the immune system have an intrinsic capacity to regulate T cell responses to haptens and that pathologic reactions occurs only when in certain circumstances the balance between effector and regulatory responses is broken.

Concluding Remarks

New insights into ACD pathomechanisms have been achieved thanks to valuable animal models and in vitro techniques which allowed a precise identification of the T cell populations involved in the immune reaction. In the next future, the major challenge will be to identify how and why effector versus regulatory T cells expand following hapten exposures, and whether in vitro or in vivo procedures could revert effector T cells into a regulatory type. The characterization of the complexity of the T cell responses to skin applied haptens, and the identification of T cell populations that function as regulators in ACD may represent a formidable tool for the development of new strategies for the immuno-modulation of cutaneous T cell mediated disorders.

References

1. Cavani A, Albanesi C, Traidl C et al. Effector and regulatory T cells in allergic contact dermatitis. Trends Immunol 2001; 22(3):118-120.
2. Krasteva M, Keheren J, Ducluzeau MT et al. Contact dermatitis I. Phatophisiology of contact sensitivity. Eur J Dermatol 1999; 9(2):144-159.
3. Grabbe S, Schwarz T. Immunoregulatory mechanisms involved in elicitation of allergic contact dermatitis. Immunol Today 1998; 19(1):37-44.
4. Robert C, Kupper TS. Inflammatory skin diseases, T cells, and immune surveillance. N Engl J Med 1999; 341(24):1817-1828.
5. Kunkel EJ, Butcher EC. Chemokines and tissue-specific migration of lymphocytes. Immunity 2002; 16(1):1-4.
6. Fuhlbrigge RC, Kieffer JD, Armerding D et al. Cutaneous lymphocyte antigen is a specialized form of PSGL-1 expressed on skin-homing T cells. Nature 1997; 389(6654):978-981.
7. Campbell JJ, Haraldsen G, Pan J et al. The chemokine receptor CCR4 in vascular recognition by cutaneous but not intestinal memory T cells. Nature 1999; 400(6746):776-780.
8. Hudak S, Hagen M, Liu Y et al. Immune surveillance and effector functions of CCR10+ skin homing T cells. J Immunol 2002; 169(3):1189-1196.
9. Sebastiani S, Albanesi C, De Pità O et al. The role of chemokines in allergic contact dermatitis. Arch Dermatol Res 2002; 293(11):552-559.
10. Grabbe S, Steinert M, Mahanke K et al. Dissection of antigenic and irritative effects of epicutaneously applied haptens in mice. Evidence that not the antigenic component but nonspecific proinflammatory effects of haptens determine the concentration-dependent elicitation of allergic contact dermatitis. J Clin Invest 1996; 98(5):1158-1164.
11. Zhang L, Tinkle SS. Chemical activation of innate and specific immunity in contact dermatitis. J Invest Dermatol 2000; 115(2):168-176.
12. Goebeler M, Gillitzer R, Kilian K et al. Multiple signaling pathways regulate NF-kB-dependent transcription of the monocyte chemoattractant protein-1 gene in primary endothelial cells. Blood 2001; 97(1):46-55.
13. Albanesi C, Scarponi C, Sebastiani S et al. A cytokine-to-chemokine axis between T lymphocytes and keratinocytes can favor Th1 cell accumulation in chronic inflammatory skin diseases. J Leukoc Biol 2001; 70(4):617-623.
14. Teunissen MB, Koomen CW, de Waal Malefyt R et al. Interleukin-17 and interferon-gamma synergize in the enhancement of proinflammatory cytokine production by human keratinocytes. J Invest Dermatol 1998; 111(4):645-649.
15. Albanesi C, Scarponi C, Cavani A et al. Interleukin-17 is produced by both Th1 and Th2 lymphocytes, and modulates interferon-γ- and interleukin-4-induced activation of human keratinocytes. J Invest Dermatol 2000; 115(1):81-87.
16. Albanesi C, Scarponi C, Sebastiani S et al. IL-4 enhances keratinocyte expression of CXCR3 agonistic chemokines. J Immunol 2000; 165(3):1395-1402.
17. Roncarolo MG, Bacchetta R, Bordignon C. Type 1 regulatory T cells. Immunol Rev 2001; 182(1):68-79.
18. Shevach EM. $CD4^+CD25^+$ suppressor T cells: More question than answers. Nat Rev Immunol 2002; 2(6):389-400.
19. Sakaguchi S, Sakaguchi N, Shimizu J et al. Immunologic tolerance maintained by $CD25^+$ $CD4^+$ regulatory T cells: Their common role in controlling autoimmunity, tumor immunity, and transplantation tolerance. Immunol Rev 2001; 182(1):18-32.
20. Kapsenberg ML, Wierenga EA, Stiekma FEM et al. Th1 lymphokine production profiles of nickel-specific $CD4^+$ T-lymphocyte clones from nickel contact allergic and nonallergic individuals. J Invest Dermatol 1992; 98(1):59-63.
21. Probst P, Kuntzlin D, Fleischer B. Th2-type infiltrating T cells in nickel-induced contact dermatitis. Cell Immunol 1995; 165(1):134-140.
22. Cavani A, Mei D, Guerra E et al. Patients with allergic contact dermatitis to nickel and non allergic individuals display different nickel-specific T cell responses. Evidence for the presence of effector $CD8^+$ and regulatory $CD4^+$ T cells. J Invest Dermatol 1998; 111(4):621-628.

23. Xu H, DiIulio NA, Fairchild RL. T cell populations primed by hapten sensitization in contact sensitivity are distiguished by polarized patterns of cytokine production: Interferon gamma-producing (Tc1) effector CD8+ T cells and interleukin (Il) 4/Il-10-producing (Th2) negative regulatory CD4+ T cells. J Exp Med 1996; 183(3):1001-1012.
24. Borg L, Christensen JM, Kristiansen J et al. Nickel-induced cytokine production from mononuclear cells in nickel-sensitive and controls. Arch Dermatol Res 2000; 292(6):285-291.
25. Werfel T, Hentschel M, Kapp A et al. Dichotomy of blood- and skin-derived IL-4-producing allergen-specific T cells and restricted Vβ repertorire in nickel-mediated contact dermatitis. J Immunol 1997; 158(5):2500-2505.
26. Kitagaki H, Ono N, Hayakawa K et al. Repeated elicitation of contact hypersensitivity induces a shift in cutaneous cytokine milieu from a T helper cell type 1 to a T helper cell type 2 profile. J Immunol 1997; 159(5):2484-2491.
27. Tang A, Judge TA, Nickoloff BJ et al. Suppression of mirine allegic contact dermatitis by CTLA4Ig. Tolerance induction of Th2 responses requires additional blockade fo CD40-ligand. J Immunol 1996; 157(1):117-125.
28. Martin S, von Bonin A, Fessler C et al. Structural complexity of antigenic determinants for class I MHC-restricted, hapten-specific T cells. J Immunol 1993; 151(2):678-687.
29. Kessler B, Michielin O, Blanchard CL et al. T cell recognition of hapten. Anatomy of T cell receptor binding of a H-2K^{d}-associated photoreactive peptide derivative. J Biol Chem 1999; 274(6):3622-3631.
30. Franco A, Yokoyama T, Huynh D et al. Fine specificity and MHC restriction of trinitrophenyl-specific CTL. J Immunol 1999; 162(6):3388-3394.
31. Vollmer J, Weltzien HU, Moulon C. TCR reactivity in human nickel allergy indicates contacts with complementary-determining region 3 but excludes superantigen-like recognition. J Immunol 1999; 163(5):2723-2731.
32. Vollmer J, Weltzien HU, Gamerdinger K et al. Antigen contacts by Ni-reactive TCR: Typical αβ chain cooperation versus a chain-dominated specificity. Int Immunol 2000; 12(12):1723-1731.
33. Sieben S, Hertl M, Al Masaouti D et al. Characterization of T cell responses to fragrances. Toxicol Appl Pharmacol 2001; 172(3):172-178.
34. Moulon C, Vollmer J, Weltzien HU. Characterization of processing requirements and metal cross-reactivities in T cell clones from patients with allergic contact dermatitis to nickel. Eur J Immunol 1995; 25(12):3308-3315.
35. Kalish RS, Wood JA, LaPorte A. Processing of urushiol (poison ivy) hapten by both endogenous and exogenous pathways for presentation to T cells in vitro. J Clin Invest 1994; 93(5):2039-2047.
36. Nasorri F, Sebastiani S, Mariani V et al. Activation of nickel-specific CD4+ T lymphocytes in the absence of professional antigen-presenting cells. J Invest Dermatol 2002; 118(6):172-179.
37. Büdinger L, Neuser N, Totzke U et al. Preferential usage of TCR-Vβ17 by peripheral and cutaneous T cells in nickel-induced contact dermatitis. J Immunol 2001; 167(10):6038-6044.
38. Vollmer J, Fritz M, Dormoy A et al. Dominance of the Vβ17 element in nickel-specific human T cell receptors relates to severity of contact sensitivity. Eur J Immunol 1997; 27(8):1865-1874.
39. Kapsemberg ML, Res P, Bos JD et al. Nickel-specific T lymphocyte clones derived from allergic nickel-contact dermatitis lesions in man: Heterogeneity based o requirement of dendritic antigen-presenting cell subsets. Eur J Immunol 1987; 17(6):861-865.
40. Dubey C, Croft M, Swain SL. Naive and effector CD4 T cells differ in their requirements for T cell receptor versus costimulatory signals. J Immunol 1996; 157(8):3280-3289.
41. Nuriya S, Enomoto S, Azuma M. The role of CTLA-4 in murine contact hypersensitivity. J Invest Dermatol 2001; 116(5):764-768.
42. Grabbe S, Steinbrink K, Steinert M et al. Removal of the majority of epidermal Langerhans cells by topical or systemic steroid application enhances the effector phase of murine contact hypersensitivity. J Immunol 1995; 155(9):4207-4217.
43. Ma W, Pober JS. Human endothelial cells effectively costimulate cytokine production by, but not differentiation of naive CD4+ T cells. J Immunol 1998; 161(5):2158-2167.
44. Marelli-Berg FM, Weetman A, Frasca L et al. Lechler RI: Antigen presentation by epithelial cells induces anergic immunoregulatory CD45RO+ T cells and deletion of CD45RA+ T cells. J Immunol 1997; 159(12):5853-5861.

45. Gaspari AA, Katz SI. Induction of in vivo hyporesponsiveness to contact allergens by hapten-modified Ia+ keratinocytes. J Immunol 1991; 147(12):4155-4161.
46. Barnaba V, Watts C, de Boer M et al. Professional presentation of antigen by activated human T cells. Eur J Immunol 1994; 24(1):71-75.
47. Arnold PY, Davidian DK, Mannie MD. Antigen presentation by T cells: T cell receptor ligation promotes antigen acquisition from professional antigen presenting cells. Eur J Immunol 1997; 27(12):3198-3205.
48. Jeannin P, Herbault N, Delneste Y et al. Human effector memory T cells express CD86: A functional role in naive T cell priming. J Immunol 1999; 162(4):2044-2048.
49. Wyllie JC, More HR, Haust MD. Electron microscopy of epidermal lesions elicited during delayed hypersensitivity. Lab Invest 1964; 13:137.
50. Kehren J, Desvignes C, Krasteva M et al. Cytotoxicity is mandatory for $CD8^+$ T cell mediated contact hypersensitivity. J Exp Med 1999; 189(5):779-786.
51. Yawalkar N, Hunger RE, Buri C et al. A comparative study of the expression of cytotoxic proteins in allergic contact dermatitis and psoriasis. Spongiotic skin lesions in allergic contact dermatitis are highly infiltrated by T cells expressing perforin and granzime B. Am J Pathol 2001; 158(3):803-808.
52. Traidl C, Sebastiani S, Albanesi C et al. Disparate cytotoxic activity of nickel-specific $CD8^+$ and $CD4^+$ T cell subsets against keratinocytes. J Immunol 2000; 165(6):3058-3064.
53. Trautmann A, Akdis M, Kleemann D et al. T cell-mediated Fas-induced keratinocyte apoptosis play a key pathogenetic role in eczematous dermatitis. J Clin Invest 2000; 106(1):25-35.
54. Matsue H, Kobayashi H, Hosokawa T et al. Keratinocytes constituively express the Fas-antigen that mediates apoptosis in IFN-gamma treated cultured keratinocytes. Arch Dermatol Res 1996; 287(3-4):315-320.
55. Sayama K, Yonehara S, Watanabe Y et al. Expression of Fas-antigen on keratinocytes in vivo and induction of apoptosis in cultured keratinocytes. J Invest Dermatol 1994; 103(3):330-340.
56. Trautmann A, Autznauer F, Akdis M et al. The differential fate of cadherins during T-cell-induced keratinocyte apoptosis leads to spongiosis in eczematous dermatitis. J Invest Dermatol 2001; 117(4):927-934.
57. Saulnier M, Huang S, Aguet M et al. Role of interferon-gamma in contact hypersensitivity assessed in interferon-gamma receptor-deficient mice. Toxicology 102(3):301-312.
58. Gocinski BL, Tigelaar RE. Roles of $CD4^+$ and $CD8^+$ cells in murine contact sensitivity revealed by in vivo monoclonal antibody depletion. J Immunol 1990; 144(11):4121-4128.
59. Bour H, Peyron E, Gaucherand M et al. Major histocompatibility complex class I-restricted $CD8^+$ T cells and class II-restricted $CD4^+$ T cells, respectively, mediate and regulate contact hypersensitivity to dinitrofluorobenzene. Eur J Immunol 1995; 25(11):3006-3010.
60. Bouloc A, Cavani A, Katz SI. Contact hypersensitivity in MHC class II-deficient mice depends on $CD8^+$ T lymphocytes primed by immunostimulating Langerhans cells. J Invest Dermatol 1998; 111(1):44-49.
61. Wang B, Fujisawa H, Zhuang L et al. CD4+ Th1 and CD8+ type 1 cytotoxic T cells both play a crucial role in the full development of contact hypersensitivity. J Immunol 2000; 165(12):6783-6790.
62. Martin S, Lappin MB, Koheler J et al. Peptide immunization indicates that CD8+ T cells are the dominant effector cells in trinitrophenyl-specific contact hypersensitivity. J Invest Dermatol 2000; 115(2):260-266.
63. Lisby S, Hansen LH, Skov L et al. Nickel-induced activation of T cells in individuals with negative patch test to nickel sulphate. Arch Dermatol Res 1999; 291(5):247-252.
64. Hauser C, Katz SI. Generation and characterization of T-helper cells by primary in vitro sensitization using langerhans cells. Immunol Rev 1990; 117:67-84.
65. Biederman T, Mailhammer R, Mai A et al. Reversal of established delayed type hypersensitivity reactions following therapy with IL-4 or antigen-specific Th2 cells. Eur J Immunol 2001; 31():1582-1591.
66. Berg DJ, Leach MW, Kuhn R et al. Interleukin 10 but not interleukin 4 is a natural suppressant of cutaneous inflammatory responses. J Exp Med 1995; 182(1):99-108.
67. Traidl C, Jugert F, Krieg T et al. Inhibition of allergic contact dermatitis to DNCB but no to oxazolone in interleukin-4 deficient mice. J Invest Dermatol 1999; 112(4):476-482.

68. Yokozeki H, Ghoreishi M, Takagawa S et al. Signal transducer and activator of transcription 6 is essential in the induction of contact hypersensitivity. J Exp Med 2000; 191(6):995-1004.
69. Hindsen M, Bruze M, Christensen OB. Individual variation in nickel patch test reactivity. Am J Contact Dermatitis 1999; 10(2):62-67.
70. Cavani A, Nasorri F, Prezzi C et al. Human CD4+ T lymphocytes with remarkable regulatory functions on dendritic cells and nickel-specific Th1 immune responses. J Invest Dermatol 2000; 14(2):295-302.
71. DiIulio NA, Xu H, Fairchild RL. Diversion of CD4+ T cell development from regulatory T helper to effector T helper cells alters the contact hypersensitivity response. Eur J Immunol 1996; 26(11):2606-2612.
72. Desvignes C, Etchart N, Kehren J et al. Oral administration of haptens inhibits in vivo induction of specific cytotoxic CD8+ T cells mediating tissue inflammation: A role for regulatory CD4+ T cells. J Immunol 2000; 164(5):2515-2522.
73. Schwartz A, Beissert S, Grosse-Heitmeyer K et al. Evidence for functional relevance of CTLA-4 in ultraviolet-radiation-induced tolerance. J Immunol 2000; 165(4):1824-1831.
74. Artik S, Haarhuis K, Wu X et al. Tolerance to nickel: Oral nickel administration induces a high frequency of anergic T cells with persistent suppressor activity. J Immunol 2001; 167(12):6794-6803.
75. Cavani A, Nasorri F, Ottaviani S et al. Cutaneous lymphocyte associated antigen (CLA)+ CD4+CD25+ cells regulate nickel-specific T cells responses. J Immunol 2003; 171:5760-5768.
76. Gregori S, Casorati M, Amuchastegui S et al. Regulatory T cells induced by 1 alpha, 25-dihydroxyvitamin D3 and mycophenolate mofetil treatment mediate transplantation tolerance. J Immunol 2001; 167(4):1945-1953.
77. Matiszak MK, Citterio S, Rescigno M et al. Differential effects of corticosteroids during different stages of dendritic cell maturation. Eur J Immunol 2000; 30(4):1233-1242.
78. Steinbrink K, Graulich E, Kubsch S et al. CD4 (+) and CD8 (+) anergic T cells induced by interleukin-10-treated human dendritic cells: Display antigen-specific suppressor activity. Blood 2002; 99(7):2468-2476.
79. Jonuleith H, Shmitt E, Steinbrink K. Dendritic cells as a tool to induce anergic and regulatory T cells. Trends Immunol 2001; 22(7):394-400.

Chapter 6

T Cell Recruitment in Allergic Contact Dermatitis

Silvia Sebastiani, Giampiero Girolomoni and Andrea Cavani

Summary

Expression of allergic contact dermatitis requires efficient recruitment of T cells and other leukocytes at the site of penetration of haptens. The ability of T cells to circulate in the skin environment is defined by their homing and chemokine receptor asset, and is directed by the chemokine repertoire expressed by resident skin cell populations, such as keratinocytes, fibroblasts, dendritic cells, mast cells and endothelial cells. Moreover, incoming T cells, once activated, become themselves a source of chemokines, and directly contribute to the afflux of new waves of circulating leukocytes. At a later time point, the recruitment of T lymphocytes with regulatory function appears critical for the termination of the immune reaction. In this scenario, the possibility to target endothelial receptors, chemokines or chemokine receptors may offer new opportunities for therapeutic interventions.

Introduction

Entry of T cells at the site of hapten challenge is a crucial event for the expression of allergic contact dermatitis (ACD), and implies complex interactions between endothelial cells and circulating T lymphocytes. Skin homing T cells account for the 20-25% of the total number of circulating T lymphocytes and can be identified thanks to the memory/effector phenotype ($CD45RA^-CD45RO^+$) and constitutive expression of the cutaneous lymphocyte associated antigen (CLA).[1-4] In addition, skin seeking T cells preferentially display certain chemokine receptors, such as CCR4 and CCR10, which contribute, together with homing receptors, to selective tissue targeting.[5-7] Expression of homing and chemokine receptors on differentiating T lymphocytes is directed by many factors, most of which still not well understood. It appears that induction of CLA on memory T cells depends on the type and origin of antigen presenting cells (APC) and on the cytokine milieu in which T cell priming occurs, being interleukin (IL)-12 the most effective inducer of the receptor.[8,9] On the other hand, it has been ascertained that the pattern of chemokine receptors expressed by T lymphocytes parallel their cytokine profile, and is modulated by the functional status of the cell.[10-14] In inflammatory conditions, skin resident cells contribute relevantly to the recruitment of T lymphocytes through the production of chemokines in a time and dose coordinated manner.[15] Activation of resident skin cells is mostly dependent on the release of interferon (IFN)-γ and tumor necrosis factor (TNF)-α by T cells. Thus the cross talk between infiltrating T lymphocytes and resident cell populations acts as a positive loop to promote the rapid and massive extravasation of circulating leukocytes

Immune Mechanisms in Allergic Contact Dermatitis, edited by Andrea Cavani and Giampiero Girolomoni. ©2005 Eurekah.com.

upon antigen encounter, and is directly responsible for the selective influx of T cell subsets with either effector or regulatory function in ACD.

Chemokines and Chemokine Receptors

Chemokines are small (6-14 kDa) secreted proteins that regulate the traffic of various types of leukocytes, including lymphocytes, dendritic cells (DCs), monocytes, neutrophils and eosinophils, by mediating their adhesion to endothelial cells, as well as their transendothelial migration and tissue invasion.[16,17] Accordingly to the position of two highly conserved cysteine residues at the N-terminus of the molecule (Table 1), chemokines are classified in the C, CC, CXC and CX_3C subfamilies. Upon secretion, chemokines bind to extracellular matrix and cell membrane proteoglycans forming stable gradients near to the site of production. CX3CL1/Fractalkine is an exception, as it is expressed as a membrane integral protein, that, however, can be shed from the membrane.[18] Although the main function of chemokines is to regulate cell trafficking within tissues, they also display important roles in leukocyte activation and differentiation.[19]

The effect of chemokines on target cells is mediated by seven transmembrane spanning, G-protein coupled receptors.[20] To date, about ten CC-, six CXC- one CX_3C- and one XC-chemokine receptors have been identified; their genes are mostly located in clusters on chromosomes 2 and 3 (Table 1). The chemokine-chemokine receptor axis is highly promiscuous, allowing a single chemokine to bind different receptors and a chemokine receptor to transduce the signal for several chemokines. In contrast to this notion, some recently identified chemokines show a very high receptor and tissue specificity, and are thought to contribute to tissue-restricted leukocyte trafficking.[6,21] In particular, CCL27/cutaneous T-cell attracting chemokine (CTACK) is predominantly expressed by keratinocytes and binds specifically CCR10.[7,22] Depending on their physiological features, including regulation and distribution of the respective receptors, chemokines can be classified as "inflammatory" and "lymphoid" or "homeostatic". Inflammatory chemokines, like CCL2/monocyte chemoattractant protein (MCP)-1, CXCL10/interferon-induced protein (IP)-10, CCL5/Regulated upon activation, normal T cell expressed and secreted (RANTES) and many others, are expressed in inflamed tissues by resident or infiltrating cells, and recruit effector cells such as monocytes, granulocytes and T cells. Homeostatic chemokines, such as CCL19/EBV-induced molecule 1 ligand chemokine (ELC)/ Macrophage inflammatory protein (MIP)-3 β, CCL21/secondary lymphoid tissue chemokine (SLC) and CXCL13/B cell-attracting chemokine (BCA)-1, are primarily produced within lymphoid tissues and are involved in the maintenance of the constitutive lymphocyte traffic and cell positioning within these organs. Several chemokines like CCL22/ macrophage-derived chemokine (MDC), CCL17/thymus- and activation-regulated chemokine (TARC), and CCL20/liver and activation-regulated chemokine (LARC) can be ascribed to both families.[5,12]

Chemokine Receptors on T Cell Subsets

Chemokine receptor asset on T lymphocytes varies depending on the state of differentiation, activation, and tissue targeting. Naive T lymphocytes mostly home to lymph nodes, where they are experienced by antigen-loaded DCs migrated from peripheral tissues. The entry of naive T cells, as well as of migrating DCs into lymph nodes is regulated by the expression of SLC that bind to CCR7.[23,24] CCR7-SLC interaction increases $\alpha_L\beta_2$ integrin adhesiveness for intercellular adhesion molecule (ICAM)-1/2 on endothelial cells, thus promoting T lymphocyte firm adhesion and transmigration. Once T cells have crossed the high endothelial venules, SLC and ELC produced by DCs and stromal cells will favor the interaction between immigrated T lymphocytes and APCs. The pivotal role of the axis CCR7/SLC and ELC in this process has been demonstrated in CCR7-deficient mice that show a reduced number of naive

Table 1. Principal human chemokines and chemokine receptors

Systematic Name	Current Name	Receptor(s)
CXC family		
CXCL1	Growth-related oncogene (GRO)α, MGSA-α, NAP-3	CXCR1, CXCR2
CXCL2	GROβ, MGSA-β, MIP-2α	CXCR2
CXCL3	GROγ, MGSA-γ, MIP-2β	CXCR2
CXCL4	PF4	Unknown
CXCL5	ENA-78	CXCR2
CXCL6	GCP-2, CKA-3	CXCR1, CXCR2
CXCL7	NAP-2	CXCR2
CXCL8	IL-8	CXCR1, CXCR2
CXCL9	Monokine-induced by interferon g (Mig)	CXCR3
CXCL10	Interferon-induced protein of 10 kDa (IP-10), C7	CXCR3
CXCL11	Interferon-induced T cell α-chemoattractant (I-TAC), IP-9	CXCR3
CXCL12	Stromal-derived factor (SDF)-1α/β	CXCR4
CXCL13	B cell activating chemokine (BCA)-1, BLC	CXCR5
CXCL14	Breast and kidney chemokine (BRAK), bolekine, MIP-2γ	Unknown
CXCL15	lungkine	Unknown
CXCL16	SR-PSOX	Unknown
CC family		
CCL1	I-309	CCR8
CCL2	Monocyte chemotactic protein (MCP)-1, MCAF, SMC-CF	CCR2
CCL3	Macrophage inflammatory protein (MIP)-1α, LD78α, SISα	CCR1, CCR5
CCL4	MIP-1β, LAG-1	CCR5
CCL5	Regulation and activated normal T cell expressed and secreted (RANTES)	CCR1, CCR3, CCR5
CCL6	Unknown	Unknown
CCL7	MCP-3	CCR1, CCR2, CCR3
CCL8	MCP-2	CCR3
CCL9/10	Unknown	Unknown
CCL11	Eotaxin-1	CCR3
CCL12	Unknown	CCR2
CCL13	MCP-4, NCC-1	CCR2, CCR3
CCL14	HCC-1, HCC-3, NCC-2	CCR1
CCL15	HCC-2, Lkn-1, MIP-1δ, NCC-3	CCR1, CCR3
CCL16	HCC-4, LEC, NCC-4	CCR1
CCL17	Thymus- and activation-regulated chemokine (TARC)	CCR4
CCL18	Dendritic cell chemokine 1 (DC-CK1), PARC, AMAC-1, MIP-4	Unknown
CCL19	MIP-3β, ELC, exodus-3	CCR7
CCL20	MIP-3α, LARC, exodus-1, ST-38	CCR6
CCL21	Secondary lymphoid tissue chemokine (SLC), 6Ckine, exodus-2	CCR7
CCL22	Macrophage-derived chemokine (MDC), STCP-1	CCR4
CCL23	MPIF-1	CCR1
CCL24	Eotaxin-2, MPIF-2	CCR3
CCL25	Thymus-expressed chemokine (TECK)	CCR9
CCL26	Eotaxin-3, MIP-4α	CCR3
CCL27	Cutaneous T cell-attracting chemokine (CTACK), ILC	CCR10
CCL28	Mucosae-associated epithelial chemokine (MEC)	CCR10

Table continued on next page

Table 1. Continued

Systematic Name	Current Name	Receptor(s)
C family		
XCL1	Lymphotactin, SCM-1α, ATAC	XCR1
XCL2	SCM-1β	XCR1
CXXXC family		
CX3CL1	Fractalkine	CX_3CR1

T cells and impaired contact hypersensitivity (CHS) responses.[24] In addition, DCs are the most abundant source of MDC and TARC that bind the CCR4 receptor expressed by recently activated T cells.[12,25] Once experienced, T lymphocytes completely rearrange their chemokine receptor profile, and acquire new migratory capacity to allow their homing in peripheral tissues. Importantly, the acquisition of discrete chemokine receptors parallel the differentiation and cytokine polarization of T cells. T helper (Th)1 lymphocytes secrete IFN-γ and express CXCR3 (the receptor for IP-10/CXCL10, Mig/CXCL9 and I-TAC/CXCL11) and CCR5 (the receptor for MIP-1β/CCL4 and RANTES/CCL5). Th2 cells release IL-4, IL-5 and IL-13, and express CCR3 (eotaxin/CCL11 receptor) and CCR8 (I-309/CCL1 receptor).[10-12] More recently, specialized subsets of CD4$^+$ T lymphocytes with regulatory functions have been described.[26-28] Among these, T regulatory (Tr)1 lymphocytes, characterized by high IL-10, but no or very low IFN-γ and IL-4 release, have been shown to be strongly involved in the modulation of ACD reactions.[29] Tr1 cells coexpress both Th1 and Th2-associated chemokine receptors, CCR7 and high levels of CCR8.[30] Interestingly, another subtype of regulatory CD4$^+$ T cells defined by the constitutive expression of CD25, specifically express the chemokine receptors CCR4 and CCR8 and respond to MDC, TARC and CCL1/I-309.[31] CCR4 and/or CCR8 may thus guide T regulatory cells to sites of antigen presentation in secondary lymphoid tissues as well as in inflamed areas, to attenuate T cell activation. Importantly, mature DCs, by secretion of the CCR4 ligands TARC and MDC, preferentially attract T regulatory lymphocytes among circulating CD4$^+$ T cells.[31] As a consequence, regulation of I-309, TARC, and MDC production during the course of inflammatory responses could dictate the extent, severity, and duration of the response by modulating recruitment of CCR4- and CCR8-bearing T regulatory cells. Chemokine receptor expression on polarized CD8$^+$ T cells slightly differs compared with that of CD4$^+$ subsets. In particular, we observed a higher expression of CXCR3 on Tc1 compared with Th1, and higher CCR4 expression in Th2 in comparison with Tc2. These differences were even more pronounced in CLA$^+$, nickel specific, CD8$^+$ and CD4$^+$ T lymphocytes.[32] Memory CD4$^+$ T lymphocytes can be also classified as central or effector memory depending on the presence or absence of CCR7 on their membrane.[33] In fact, central memory, but not effector memory T cells, maintain the capacity to upregulate CCR7 and migrate to lymph nodes upon activation. Here, they can provide strong help for DC maturation and than differentiate into effector cells. A similar pool of cells able to avidly home to lymphoid organs through CCR7 and its ligands and moderately to sites of inflammation is also present within the CD8$^+$ population.[34]

T cell activation strongly affects chemokine receptor expression. In fact, the majority of receptors for inflammatory chemokines are downregulated, with the exception of CCR4 and CCR8 that are transiently upregulated.[10,14,30] Receptor down-regulation and desensitization promotes a switch from a migratory to a stationary behavior thus ensuring the persistence of immune effector cells at the site of antigen presentation.

Skin Homing T Cells

The existence of subsets of memory T cells that preferentially migrate to the skin is well documented. Tissue targeting is achieved by regulated expression of particular homing receptors on lymphocytes, and their counter receptors on endothelial, epithelial and inflammatory cells in the target tissue. Circulating lymphocytes associated with skin homing express the CLA, a sialyl lewis X-like carbohydrate epitope displayed on the P-selectin glycoprotein ligand 1 (PSGL-1) molecule.[1-3] The posttranslational modification of PSGL-1 is regulated by several glycosyltransferases, including 1,3-fucosyltransferase (FucT) IV and FucTVII, and β1,4-galactosyltransferase I. CLA expression on T cells correlates with their capacity to roll over and adhere to E-selectin exposed on activated skin microvasculature. The vast majority of CLA^+ skin-seeking $CD4^+$ T lymphocytes displays CCR4 independently from the cytokine profile. This receptor permits T lymphocyte adhesion to TARC, which is exposed on the surface of activated endothelium in inflamed skin.[6] CCR4 triggering by TARC induces T lymphocyte integrin activation and promotes the firm adhesion of circulating lymphocytes to the endothelium. More recently it has been shown that another chemokine-chemokine receptor pair, represented by CTACK and its receptor CCR10, provides a specific signal for the recruitment of CLA^+ memory T cells both in homeostatic and inflammatory skin conditions.[7,22] In fact most of the lymphocytes infiltrating the skin in psoriasis, atopic dermatitis and allergic contact dermatitis are $CCR10^+$. Moreover, its ligand, CTACK, is expressed by basal keratinocytes in steady state and is upergulated by proinflammatory stimuli like TNF-α and IL-1β. Besides being expressed in the epidermis, CTACK is also present on endothelial cells, where cooperates with TARC in inducing firm adhesion of lymphocytes and the initiation of their transendothelial migration. Thanks to its localization, CTACK together with CXL9/Mig, IP-10 and I-TAC produced by resident cells, sustains the gradient that drives skin infiltrating T cells from the perivascular area to a subepidermal location. By now, CCR10 represents a very selective marker for T cells recruitment in to the skin, although expressed by only 30-40% of CLA^+ T cells.

Finally, CCR6, albeit also expressed in gut migrating lymphocytes, has been described as an important chemokine receptor that favors T cell migration into the skin.[35-37] In particular, TNF-α-activated dermal endothelial cells, produce large amounts of MIP-3α which is critical for arrest of $CCR6^+$ memory T cells. In addition, MIP-3α produced by keratinocytes and endothelial cells has been also suggested to play a role in the selective recruitment of Langerhans cells in the skin.[38]

The Control of T Cell Migration in ACD

The inflammatory infiltrate in ACD is composed mainly of $CD4^+$ and $CD8^+$ T cells, monocytes and DCs, with an early and transient presence of neutrophils.[39] The expression of both murine CHS and human ACD correlates with the activity of hapten-specific $CD8^+$ T cells, which exert their effector function trough direct cytotoxic activity as well as the release of cytokines.[40-44] On the other hand, $CD4^+$ T cells play a more complex role, with Th1 and Th2 subsets contributing to disease expression, and T regulatory cells primarily involved in its modulation and/or termination.[39] In addition, Th2 cells may also exert a regulatory role as IL-4 administration or passive transfer of hapten-specific Th2 lymphocytes can reverse established murine CHS.[45] However, CHS is reduced or unmodified in $IL\text{-}4^{-/-}$ mice and decreased in STAT-6 deficient mice, in which Th2 responses are impaired.[46]

During ACD, leukocyte recruitment is under the control of the sequential and coordinated release of chemokines from resident and immigrating cells. The expression pattern of chemokines in the skin at different time points after hapten application has been recently described.[47] By using in situ hybridization, it has been shown that MCP-1 is already detectable at 6 h after hapten challenge, whereas RANTES and MDC appear at 12 h concomitantly with the infiltration of mononuclear cells in the dermis and epidermis. The expression of IP-10, Mig, TARC and PARC began at 12 h and peaked at 72 h, paralleling the strong infiltration of lymphocytes. The variety of chemokines expressed during ACD determines a robust and rapid recruitment of leukocytes into the skin. MCP-1 seems to play a relevant role in ACD. In fact, transgenic mice overexpressing MCP-1 in basal keratinocytes showed enhanced CH responses together with an increased number of infiltrating DCs.[48] It is still unclear which chemokine(s) drives the initial influx of T cells, the cellular source(s) as well as the nature of the induction stimulus. Keratinocytes may represent important contributors to this phase, although a TNF-α-mediated induction by haptens has been demonstrated only for CXCL8/IL-8.[49] Activated endothelial cells can conceivably contribute to early leukocyte arrival in the skin, by expressing both adhesion molecules and chemokines.[3,5,6,50] Also mast cells and platelets might be indirectly involved in this early T cell recruitment through the release of serotonin, which together with TNF-α activate the endothelium facilitating cell entrance. This phenomenon seems to be C5a-dependent, since C5a knockout mice do not show T cell infiltration at the sites of hapten challenge.[51] Moreover, in murine CHS, mast cells can be importantly involved in neutrophil recruitment trough the release of TNF-α and MIP-2, the functional analogue of human IL-8.[52] Infiltrating monocytes and DCs are themselves a source of chemokines for successive boosts of lymphocyte arrival. In particular several investigations demonstrated that DC-derived MDC induces the binding of antigen primed T lymphocytes to activated DCs through CCR4 receptor and that the MDC-CCR4 system plays an important role in the formation of T cell-DC clusters in inflamed skin as well as in secondary lymphoid tissue.[53,54] The activation of some hapten-specific T lymphocytes into the skin, leads to the production of cytokines like IFN-γ, TNF-α and IL-4, which in turn stimulate keratinocytes and other cells to produce IP-10, Mig and I-TAC, the ligands for CXCR3.[15,55,56] Differently from MCP-1, MDC and RANTES, these chemokines selectively attract T lymphocytes, which then rapidly accumulate in the epidermis. Indeed keratinocytes, continuously stimulated by T cell-derived cytokines produce large amounts of CXCR3 ligands thus contributing to further accumulation of CXCR3-bearing T cells. The result is that more than 70% of T lymphocytes infiltrating established ACD are $CXCR3^+$.[15,57] Comparative analysis on chemokine receptor expression on $CD4^+$ and $CD8^+$ nickel-specific T lymphocytes revealed a reciprocal higher expression of CXCR3 on $CD8^+$ and CCR4 on $CD4^+$ T cells, suggesting that trafficking of $CD4^+$ and $CD8^+$ T lymphocytes is directed primarily by the CCR4/TARC and CXCR3/IP-10 axis, respectively.[32] Once recruited into the skin, activated T lymphocytes represent a relevant source of chemokines. Upon T cell receptor triggering, Th1 and Th2 cells produce similar sets of chemokines, including RANTES, I-309, IL-8 and MDC.[32] Some differences seem to exist in terms of chemokine production between $CD4^+$ and $CD8^+$ T lymphocytes. In a mouse model it has been shown that IP-10 expression is primarily mediated by $CD8^+$ and inhibited by $CD4^+$ T lymphocytes during the elicitation phase of CHS.[58]

Resolution of ACD is likely due to multiple mechanisms including induction of T cell anergy and active suppression, and may involve several cell types. In this process IL-10 producing Tr1 cells might play a central role.[29,30,59] IL-10 producing cells migrate in vitro in response to various chemokines including MCP-1, MIPs and TARC. More interestingly, I-309, which is not active on Th1 cells, attracts more vigorously IL-10 producing lymphocytes than Th2 cells. Consistent with these results, IL-10 producing cells express higher levels of CCR8 compared to Th2 cells. I-309 is produced by both keratinocytes and activated T cells, and is expressed in ACD skin with an earlier kinetics compared to IL-4 and IL-10.[30] These data indicate that I-309/CCR8 may contribute relevantly to the termination of ACD through the recruitment of IL-10 producing lymphocytes. Also CCR6, the receptor of MIP-3α, seems to be involved in the modulation of CHS; in fact, CCR6 deficient mice show increased and persistent CHS responses, probably in relation to an impaired recruitment of $CD4^+$ regulatory cells.[60]

Concluding Remarks

Chemokines appear crucial regulators of both the induction and expression of ACD. The kinetics and pattern of chemokine expression during ACD indicate an early involvement of IL-8, followed by MCP-1 and RANTES, and by a robust burst of CXC3 receptor agonists. Lately, I-309 is produced by activated keratinocytes, DCs and infiltrating T cells. Although certain chemokines (IP-10, IP-9 and MIG) have been indicated to differentiate ACD from irritant contact dermatitis,[57] chemokine expression observed in ACD closely resemble that observed in other skin inflammaotry conditions suggesting that the skin sets up a standard sequential pattern of chemokine expression in response to different types of injuries.[61]

The rapidly growing knowledge on the mechanisms controlling the recruitment of T cells in the skin during human ACD, will permit the design of new therapies targeting adhesion molecules, chemokines or their receptors.[62-64] This possibility is only limited by the redundancy of the chemokine system: when a chemokine is blocked another takes over. In the case of delayed-type hypersensitivity reaction, for example, lymphocyte recruitment is completely inhibited only when, both CCR4 and CCR10 are blocked. Nevertheless, thanks to the mouse model, this disorder can provide an excellent opportunity for testing candidate drugs.

References

1. Picker LJ, Michie SA, Rott LS et al. A unique phenotype of skin-associated lymphocytes in humans. Preferential expression of the HECA-452 epitope by benign and malignant T cells at cutaneous sites. Amer J Path 1990; 136(6):1053-1068.
2. Akdis M, Klunker S, Schliz M et al. Expression cutaneous lymphocyte-associated antigen on human CD4+ and CD8+ Th2 cells. Eur J Immunol 2000; 30(12):3533-3541.
3. Robert C, Kupper TS. Inflammatory skin diseases, T cells, and immune surveillance. N Engl J Med 1999; 341(24):1817-1828.
4. Soler D, Humpreys TL, Spinola SM et al. CCR4 versus CCR10 in human cutaneous Th lymphocyte trafficking. Blood 2003; 101:1677-1682.
5. Moser B, Loetscher P. Lymphocyte traffic control by chemokines. Nat Immunol 2001; 2(2):123-128.
6. Campbell JJ, Haraldsen G, Pan J et al. The chemokine receptor CCR4 in vascular recognition by cutaneous but not intestinal memory T cells. Nature 1999; 400(6746):776-780.
7. Homey B, Alenius H, Müller H et al. CCL27-CCR10 interactions regulate T cell-mediated skin inflammation. Nature Med 2002; 8(2):157-165.
8. Wagers AJ, Waters CM, Stoolman LM et al. Interleukin 12 and interleukin 4 control T cell adhesion to endothelial selectins through opposite effects on a 1,3-fucosyltransferase VII gene expression. J Exp Med 1998; 188(12):2225-2231.
9. Leung DY, Gately M, Trumble A et al. Bacterial superantigens induce T cell expression of the skin-selective homing receptor, the cutaneous lymphocyte-associated antigen, via stimulation of interleukin 12 production. J Exp Med 1995; 181(2):747-753.
10. D'Ambrosio D, Iellem A, Bonecchi R et al. Selective up-regulation of chemokine receptors CCR4 and CCR8 upon activation of polarized human type 2 Th cells. J Immunol 1998; 161(10):5111-5115.
11. Bonecchi R, Bianchi G, Bordignon PP et al. Differential expression of chemokine receptors and chemotactic responsiveness of type 1 T helper cells (Th1s) and Th2s. J Exp Med 1998; 187(1):129-134.
12. Sallusto F, Mackay CF, Lanzavecchia A. The role of chemokine receptors in primary, effector, and memory immune responses. Annu Rev Immunol 2000; 18:593-620.
13. Weninger W, Crowley MA, Manjunath N et al. Migratory properties of naive, effector, and memory CD8+ T cells. J Exp Med 2001; 194(7):953-956.
14. Sallusto F, Kremmer E, Palermo B et al. Switch in chemokine receptor expression upon TCR stimulation reveals novel homing potential for recently activated T cells. Eur J Immunol 1999; 29(6):2037-204.
15. Albanesi C, Scarponi C, Sebastiani S et al. A cytokine-to-chemokine axis between T lymphocytes and keratinocytes can favor Th1 cell accumulation in chronic inflammatory skin diseases. J Leukoc Biol 2201; 70(4):617-623.

16. Mackay CR. Chemokines: Immunology's high impact factor. Nat Immunol 2001; 2(2):95-101.
17. Zlotnik A, Yoshie O. Chemokines: A new classification system and their role in immunity. Immunity 2000; 12(2):121-127.
18. Bazan JF, Bacon KB, Hardiman G et al. A new class of membrane-bound chemokine with a CX3C motif. Nature 1997; 385(6617):640-644.
19. Luther SA, Cyster J. Chemokines as regulators of T cell differentiation. Nat Immunol 2001; 2(2):102-107.
20. Rossi D, Zlotnik A. The biology of chemokines and their receptors. Annu Rev Immunol 2000; 18:217-242.
21. Zabel BA, Agace WW, Campbell JJ et al. Human G protein-coupled receptor GPR9-6/CC chemokine receptor 9 is selectively expressed on intestinal homing T lymphocytes, mucosal lymphocytes, and thymocytes and is required for thymus-expressed chemokine-mediated chemotaxis. J Exp Med 1999; 190(9):1241-1256.
22. Homey B, Wang W, Soto H et al. The orphan chemokine receptor G protein-coupled receptor-2 (GPR-2, CCR10) binds the skin associated chemokine CCL27 (CTACK/ALP/ILC). J Immunol 2000; 164(7):3465-3470.
23. Gunn MD, Tangemann K, Tam C et al. A chemokine expressed in lymphoid high endothelial venules promotes the adhesion and chemotaxis of naive T lymphocytes. Proc Natl Acad Sci USA 1998; 95(1):258-263.
24. Förster R, Schubel A, Breitfeld D et al. CCR7 coordinates the primary immune response by establishing functional microenvironments in secondary lymphoid organs. Cell 1999; 99(1):23-33.
25. Vulcano M, Albanesi C, Stopacciaro A et al. Dendritic cells as a major source of macrophage-derived chemokine/CCL22 in vitro and in vivo. Eur J Immunol 2001; 31(3):812-822.
26. Roncarolo MG, Bacchetta R, Bordignon C. Type 1 regulatory T cells. Immunol Rev 2001; 182(1):68-79.
27. Shevach EM. $CD4^+CD25^+$ suppressor T cells: More question than answers. Nat Rev Immunol 2002; 2(6):389-400.
28. Sakaguchi S, Sakaguchi N, Shimizu J et al. Immunologic tolerance maintained by $CD25^+$ $CD4^+$ regulatory T cells: Their common role in controlling autoimmunity, tumor immunity, and transplantation tolerance. Immunol Rev 2001; 182(1):18-32.
29. Cavani A, Nasorri F, Prezzi C et al. Human $CD4^+$ T lymphocytes with remarkable regulatory functions on dendritic cells and nickel-specific Th1 immune responses. J Invest Dermatol 2000; 114(2):295-302.
30. Sebastiani S, Allavena P, Albanesi C et al. Chemokine receptor expression and function in $CD4^+$ T lymphocytes with regulatory activity. J Immunol 2001; 116(2):996-1002.
31. Iellem A, Mariani M, Lang R et al. Unique chemotactic response profile and specific expression of chemokine receptors CCR4 and CCR8 by CD4(+)CD25(+) regulatory T cells. J Exp Med 2001; 194(6):847-53.
32. Sebastiani S, Albanesi C, Nasorri F et al. Nickel-specific $CD4^+$ and $CD8^+$ T cells display distinct migratory responses to chemokines produced during allergic contact dermatitis. J Invest Dermatol 2002; 118(6):1052-1058.
33. Sallusto F, Lenig D, Forster R et al. Two subsets of memory T lymphocytes with distinct homing potential and effector function. Nature 1999; 401(6754):708-712.
34. Weninger W, Crowley MA, Manjunath N et al. Migratory properties of naive, effector, and memory CD8+ T cells. J Exp Med 2001; 194(7):953-956.
35. Liao F, Rabin RL, Smith CS et al. CC-chemokine receptor 6 is expressed in diverse memory subsets of T cells and determines responsiveness to macrophage inflammatory protein 3α. J Immunol 1999; 162(1):186-194.
36. Fitzhugh DJ, Naik S, Caughman SW et al. C-C chemokine receptor 6 is essential for arrest of a subset of memory T cells on activated dermal microvascular endothelial cells under physiologic flow conditions in vitro. J Immunol 2000; 165(12):6677-6681.
37. Homey B, Dieu-Nosejan MC, Wiesenborn A et al. Up-regulation of macrophage inflammatory protein 3α/CCL20 and CC chemokine receptor 6 in psoriasis. J Immunol 2000; 164(12):6621-6632.
38. Charbonnier AS, Kohrgruber N, Kriehuber E et al. Macrophage inflammatory protein 3α is involved in the constitutive trafficking of epidermal Langerhans cells. J Exp Med 1999; 190(12):1755-1767.

39. Cavani A, Albanesi C, Traidl C et al. Effector and regulatory T cells in allergic contact dermatitis. Trends Immunol 2001; 22(3):118-120.
40. Bouloc A, Cavani A, Katz SI. Contact hypersensitivity in MHC class II-deficient mice depends on $CD8^+$ T lymphocytes primed by immunostimulating Langerhans cells. J Invest Dermatol 1998; 111(1):44-49.
41. Kehren J, Desvignes C, Krasteva M et al. Cytotoxicity is mandatory for $CD8^+$ T cell-mediated contact hypersensitivity. J Exp Med 1999; 189(5):779-786.
42. Wang B, Fujisawa H, Zhuang L et al. $CD4^+$ Th1 and $CD8^+$ type 1 cytotoxic T cells both play a crucial role in the full development of contact hypersensitivity. J Immunol 2000; 165(12):6783-6790.
43. Traidl C, Sebastiani S, Albanesi C et al. Disparate cytotoxic activity of nickel-specific $CD8^+$ and $CD4^+$ T cell subsets against keratinocytes. J Immunol 2000; 165(6):3058-3064.
44. Trautmann A, Akdis M, Kleemann D et al. T cell-mediated Fas-induced keratinocyte apoptosis plays a key pathogenetic role in eczematous dermatitis. J Clin Invest 2000; 106(1):9-10.
45. Biedemann T, Mailhammer R, Mai A et al. Reversal of established delayed type hypersensitivity reactions following therapy with IL-4 or antigen-specific Th2 cells. Eur J Immunol 2001; 31(5):1582-1591.
46. Yokozeki H, Ghoreishi M, Takagawa S et al. Signal transducer and activator of transcription 6 is essential in the induction of contact hypersensitivity. J Exp Med 2000; 191(6):995-1004.
47. Goebeler M, Trautmann A, Voss A et al. Differential and sequential expression of multiple chemokines during elicitation of allergic contact hypersensitivity. Am J Pathol 2001; 158(2):431-440.
48. Nakamura K, Williams IR, Kupper TS. Keratinocyte-derived monocyte chemoattractant protein 1 (MCP-1): Analysis in a transgenic model demonstrates MCP-1 can recruit dendritic and Langerhans cells to skin. J Invest Dermatol 1995; 105(95):635-643.
49. Griffiths CE, Barker JN, Kunkel S. Modulation of leukocyte adhesion molecule, a T-cell chemotaxin (IL-8) and a regulatory cytokine (TNFα) in allergic contact dermatitis (rhus dermatitis). Br J Dermatol 1991; 124(6):519-531.
50. von Andrian UH, Mackay CR. T-cell function and migration. Two sides of the same coin. N Engl J Med 2001; 343(14):1020-1034.
51. Tsuji RF, Kawikova I, Ramabhadran R et al. Early local generation of C5a initiates the elicitation of contact sensitivity by leading to early T cell recruitment. J Immunol 2000; 165(3):1588-1598.
52. Biedermann T, Kneilling M, Mailhammer R et al. Mast cells control neutrophil recruitment during T cell-mediated delayed-type hypersensitivity reactions through tumor necrosis factor and macrophage inflammatory protein 2. J Exp Med 2000; 192(10):1441-1451.
53. Wu M, Fang H, Hwang ST. CCR4 mediates antigen primed T cell binding to activated dendritic cells. J Immunol 2001; 167(9):4791-4795.
54. Katou F, Ohtani H, Natayama T et al. Macrophage-derived chemokine (MDC/CCL22) and CCR4 are involved in the formation of T lymphocyte-dendritic cell clusters in human inflamed skin and secondary lymphoid tissue. Am J Pathol 2001; 158(4):1263-1270.
55. Albanesi C, Cavani A, Girolomoni G. Interleukin-17 is produced by nickel-specific T lymphocytes and regulates ICAM-1 expression and chemokine production in human keratinocytes synergistic or antagonist effects with interferon-γ and tumor necrosis factor-α. J Immunol 1999; 162(1):494-502.
56. Albanesi C, Scarponi C, Sebastiani S et al. IL-4 enhances keratinocyte expression of CXCR3 agonistic chemokines. J Immunol 2000; 165(3):1395-1402.
57. Flier J, Boorsma DM, Bruynzeel DP et al. The CXCR3 activating chemokines IP-10 Mig, and IP-9 are expressed in allergic but non in irritant patch test reactions. J Invest Dermatol 1999; 113(4):574-578.
58. Abe M, Kondo T, Xu H et al. Interferon-γ inducible protein (IP-10) expression is mediated by $CD8^+$ T cells and is regulated by $CD4^+$ T cells during the elicitation phase of contact hypersensitivity. J Invest Dermatol 1996; 107(3):360-366.
59. Schwarz A, Beissert S, Grosse-Heitmeyer K et al. Evidence for functional relevance of CTLA-4 in ultraviolet-radiation-induced tolerance. J Immunol 2000; 165(4):1824-1831.
60. Varona R, Villares R, Carramolino L et al. CCR6-deficient mice have impaired leukocyte homeostasis and altered contact hypersensitivity and delayed-type hypersensitivity responses. J Clin Invest 2001; 107(6):R37-R45.

61. Engelhardt E, Toksoy A, Goebeler M et al. Chemokines IL-8, Gro-α, MCP-1, IP-10, and MIG are sequentially and differentially expressed during phase-specific infiltration of leukocyte subset in human wound healing. Am J Pathol 1998; 153(6):1849-1860.
62. Schön MP, Krahn T, Schön M et al. Efomycine M, a new specific inhibitor od selectin, impairs leukocyte adhesion and alleviates cutaneous inflammation. Nat Medicine 2002; 8(4):336-372.
63. Zollner TM, Podda M, Pien C et al. Proteasome inhibition reduces superantigen-mediated T cell activation and the severity of psoriasis in a SCID-hu model. J Clin Invest 2002; 109(5):671-679.
64. Carter PH. Chemokine receptor antagonism as an approach to anti-inflammatory therapy: "Just right" or plain wrong? Curr Opin Chem Biol 2002; 6:510-525.

Chapter 7

Cytokines in Contact Sensitivity

Alexander H. Enk

Summary

Allergic contact dermatitis (ACD) is one of the best-established model diseases for T cell-mediated immune responses. Besides the important role that T cells play in the initiation and maintenance of this immune response, cytokines have been shown to be important mediators of the inflammatory process. Some of them, such as interleukin (IL)-1β and tumor necrosis factor (TNF)-α, are essential mediators in the induction process of ACD. The lack of these factors resulted in the failure to initiate ACD reactions. Others, such as IL-12 or IL-18, are important factors in the differentiation of the T cell response, governing the development of interferon (IFN)-γ-producing $CD8^+$ T cells as major effector cells for ACD. IL-10 or cytokine synthesis inhibitory factor has been shown to be an important regulatory factor for ACD. Production of IL-10 at later stages of ACD-development has been linked to couterregulatory processes resulting in the induction of e.g., regulatory T cells. It will be the goal of this chapter to summarize and evaluate current knowledge on the role of these cytokines in ACD.

Introduction

Besides the induction of certain MAP-kinases by contact sensitizers, the induction of a whole cascade of cytokines early after application of a contact allergen to human or murine skin has been reported as one of the initial events leading to sensitization.[1] In fact, basically all cytokines known to be produced in epidermal and dermal tissues are induced by the application at least of a strong hapten. Thus, there are numerous reports on the induction of IL-1α, IL-1β, TNF-α, IL-4, IL-6, IL-8, IL-10, IL-12, IL-15, IL-16, IL-18, as well as various chemokines and growth factors by the application of a hapten to skin.[2] Some of these studies have focused on the induction of mRNA-signals, others have also shown protein production. This plethora of cytokines has made it difficult to distinguish between essential mediators of ACD whose absence results in a failure to respond to haptens or at least in a distorted reaction to haptens, and cytokines that are merely bystanders that promote nonspecific inflammatory reactions (which is certainly also of importance). The availability of transgenic animals has largely enhanced the possibilities to dissect the importance of individual cytokines as mediators of ACD. This chapter will focus on essential mediators of ACD and review the current knowledge of their importance.

Immune Mechanisms in Allergic Contact Dermatitis, edited by Andrea Cavani and Giampiero Girolomoni. ©2005 Eurekah.com.

Interleukin-1β

One of the first cytokines to be recognized as an essential mediator of ACD was IL-1β.[3,4] Initial time course experiments were performed on murine skin using a semiquantitative RT-PCR assay. It was shown that following application of a contact allergen, IL-1β mRNA signals were the first cytokine signals that were detectable as early as 15 min after hapten treatment. In cell-depletion experiments using complement-mediated cytolysis, epidermal Langerhans cells (LC) were shown to be the prime sources of this cytokine in the epidermis.[1] These mRNA-data were confirmed by protein analysis. Further experiments revealed that neutralization of IL-1β activity in the induction phase of ACD by injection of neutralizing mAb into skin was able to completely block the induction of sensitization in mice. Importantly, neutralizing mAb for IL-1α (a product of epidermal keratinocytes) did not block sensitization. These experiments were confirmed by injection studies using recombinant IL-1β, or IL-1α. These studies showed that injection of IL-1β, but not IL-1α mimicked the phenotypic and functional changes induced by the epicutaneous application of a contact sensitizer. Only injection of IL-1β, but not injection of the respective controls caused upregulation of MHC class II molecules and migration in LC in situ. Furthermore, LC derived from IL-1β injected skin (similar to LC derived from hapten-treated skin) were superior inducers of T cell proliferation as compared to LC derived from IL-1α-injected skin. Also, IL-1β injection caused a more complete reduction of E-Cadherin-expression on the surface of LC than IL-1α thus facilitating LC migration from the epidermis.

These initial studies were confirmed by experiments performed with transgenic animals that lack expression of IL-1β. Here it could be demonstrated that IL-1β-deficient mice show impaired contact sensitivity responses to certain haptens. Sensitization was restored either by injection of IL-1β into skin prior to hapten application of by using high doses of haptens (thereby probably bypassing the epidermal compartment). Interestingly, impaired immune responses seemed to be restricted to certain haptens, while immune responses to other haptens proved to be completely normal.

As a further confirmation of the importance of IL-1 for the induction of ACD, injection or overexpression of IL-1-receptor antagonist (IL-1RA) in the epidermis resulted in a profound blockade of epidermal sensitization.

In addition, the essential role of IL-1β for the initiation of ACD was supported by analysis of caspase-1-deficient mice. LC in these mice show deficient migratory behaviour following application of a contact allergen. Injection of IL-1β can completely restore the migration of LC and thus enable sensitization to haptens in these mice.

It should be mentioned at this point, that two other studies using separate strains of IL-1β-deficient mice did not fully support the essential role of IL-1β for the initiation of ACD. Zheng et al did not observe an inhibition of ACD reactions in IL-1β knock out (KO) animals to the hapten oxazolone (OX).[6] Most of the other studies had used trinitrochlorobenzene (TNCB) or dinitrofluorobenzene (DNFB) for their analysis. It is thus likely that various haptens show varying degrees of dependency on IL-1β signals for initiation of immune reactivity. Using separate strains of IL-1β and IL-1α KO mice Nakae et al showed that IL-1α, but not IL-1β was essential for the induction of ACD in their system.[7] These authors used a different genetic background for their mice than the other studies which might explain the differential results.

In aggregate the data presented support the notion that IL-1β is indeed one of the earliest mediators essential for the induction of ACD in mice. At least some mRNA-studies in humans seem to support these findings.

Tumor Necrosis Factor-α

As a second factor that was shown to be essential for the induction phase of ACD, tumor necrosis factor (TNF)-α was identified. TNF is a product of activated keratinocytes in the epidermis, while LC seem to be less important in its production. Following application of a hapten to the epidermis, TNF induction is rapid, but occurs somewhat later than induction of IL-1β. In fact, TNF mRNA signals were enhanced within 30min of epicutaneous hapten application.[1] Thus IL-1β and TNF-α are the first cytokines to be induced by hapten application epicutaneously.

The essential function of TNF for epicutaneous sensitization was further corroborated by the finding that neutralizing mAb for TNF prevented epicutaneous sensitization, as well as the elicitation of ACD in already sensitized animals.[8] Also, injection of TNF-α into skin mimicked the phenotypic and functional changes induced by the epicutaneous application of a contact sensitizer such as upregulation of MHC class II molecules, downregulation of E-cadherin expression and enhanced accessory activity.[9]

Further studies were performed in mice lacking TNF-α expression. Here, application of the hapten OX did not result in sensitization when tested epicutaneously on TNF-deficient mice.[10] In support of these findings, analysis of TNF-Rp75-deficient mice revealed that these animals show a reduction of contact hypersensitivity (CHS) responses to epicutaneous application of hapten.[11] As an explanation a reduced migration of epidermal LC was observed in these animals following hapten painting. In contrast, no reduction of CHS reactions, but in fact epidermal hyperresponsiveness were observed in TNF-Rp55 KO mice.[12] Here enhanced ear swelling reactions were observed in mice following epicutaneous hapten application. Also LC migration was normal in these animals. In addition, injection of TNF-α into the skin of TNF-Rp55 KO mice induced significant migration of LC into the regional lymph node, while it had no effect on the migratory capacity of LC in TNF-Rp75 KO mice. No definite explanation is given for these findings, especially as the TNF-Rp55 is considered to be the more important signalling receptor for TNF, and as the TNF-Rp55 is the only receptor expressed on human and murine keratinocytes.

Recent experiments with mast cell-deficient mice have shed some new light on the importance of TNF for epicutaneous sensitization.[13] Mast cell-deficient mice show a lack of epidermal reactivity following hapten application. This lack of epidermal reactivity is due to a defect capacity to recruit PMN into the tissue. This reactivity can be restored by substitution of bone-marrow-derived mast cell precursors from wild-type, but not TNF KO mice. These findings support the notion that mast cell-derived TNF plays an important role in the recruitment of PMN into the tissue and thereby governs skin inflammation.

In aggregate, these findings indicate that TNF-α is another essential mediator of initial and ongoing ACD reactions not only for the induction of LC migration and differentiation, but also for the recruitment of nonspecific inflammatory cells into the tissue.

Interleukin-12

IL-12 is a 70-kDa heterodimeric cytokine comprised of covalently linked 35-kDa and 40 kDa subunits. Interleukin-12 is one of the dominant factors responsible for the development of Th1/Tc1-like immune reactions. As ACD is now considered to be a Tc1-type immune response, a role for IL-12 in its induction appears likely. In fact, release of IL-12 by human keratinocytes following epicutaneous application of a contact sensitizer has been described several years ago.[14] Remarkably, only contact sensitizers, but not irritants were able to induce the cytokine in human keratinocyte cultures.

Further experiments were performed in mice. Here several investigators demonstrated, that IL-12 was detectable not in skin, but in regional lymph nodes following application of contact allergens. Whereas IL-12 p35 chains are expressed constitutively in lymph node tissue, IL-12 p40 is induced within 12-24h following the epicutaneous application of a contact allergen.[15] These initial mRNA findings were confirmed by protein analysis of the biologically active IL-12 from culture supernatants of lymph node cells. In addition, double-label in situ hybridization studies revealed dendritic cells (DC) as major sources of IL-12 in the lymph node. Antibody neutralization studies revealed, that blockade of IL-12 prevented sensitization in hapten-treated mice. On the other hand, injection of IL-12 into UV-tolerized animals prevented tolerance induction in mice.

These studies were further corroborated by findings that neutralization of IL-12 in the induction phase of CHS not only prevented the epicutaneous sensitization of mice, but also induced allergen-specific tolerance.[16] Also, injection of anti-IL-12 mAb in already sensitized animals prior to hapten challenge significantly reduced ear swelling reactions. These latter findings suggest a role for IL-12 also in the elicitation phase of ACD.

Several further studies have focused on the molecular and cellular targets of IL-12 action in ACD. It was demonstrated that IL-12 reverses established tolerance against contact allergens in a high-dose intravenous tolerance model by upregulating the expression of CD80 and CD86.[17] Reversal of tolerance by IL-12 could be blocked by neutralizing mAb against CD80, and, more effectively, CD86. It thus appears that IL-12 upregulates costimulation and fosters T cell proliferation.

Other investigators have reported on a preferential activity of IL-12 on $CD8^+$ T cells.[18,19] It was demonstrated that IL-12-treatment of mice during hapten sensitization augmented IFN-γ production in $CD8^+$ T cells even in the absense of a CD40 signal from the corresponding APC (as shown by blocking experiments). In contrast, development of $CD4^+$ T cells following IL-12 injection was strictly dependent on CD40 ligation. Thus IL-12 seems to preferentially augment $CD8^+$ T cell responses in ACD even in the absence of a CD40 signal.

In aggregate the data demonstrate that IL-12 is one of the essential factors governing the quality of the immune response characteristic for ACD, namels a Tc1-response carried largely by $CD8^+$ T cells.

Interleukin-18

Interleukin-18 is a recent entry into the field of cytokines important for the development of ACD. IL-18 belongs to the family of IL-1-like cytokines, but resembles in function more closely IL-12. As IL-12, derived from DC, was shown to be a crucial mediator for ACD as outlined above, IL-18, due to its IFN-γ-inducing capabilities was investigated. Importantly, IL-18 had previously been shown to be another DC-derived cytokine.

Studies in the initiation phase of ACD demonstrated that IL-18 mRNA was significantly upregulated in the skin-draining lymph nodes.[20] Migratory, hapten-modified LC in draining lymph-nodes expressed high levels of IL-18 mRNA as detected by in situ hybridization. In addition, secreted functional IL-18 could be detected in the supernatants of hapten-stimulated lymph-node cells. Production of IFN-γ by lymph node cells following in vitro stimulation with IL-12 could be partially reversed by anti-IL-18 mAb, suggesting a synergistic role for endogenous IL-18 in IFN-γ production by lymph-node cells. Further in vivo studies showed that ACD responses were severely impeded in mice treated with neutralizing IL-18 mAb during the induction phase of contact sensitivity reactions.

These findings were supported by analysis of caspase-1-deficient animals.[21] Caspase-1 is required for functional activation of IL-18 precursor proteins. In agreement with an important function of IL-18 in ACD, caspase-1 deficient mice showed reduced swelling reaction following hapten application.

In aggregate the data support the notion that IL-18 contributes to the Tc1-twist in immune responses characteristic for ACD.

Interleukin-10

Interleukin-10, originally termed cytokine-synthesis-inhibitory factor (CSIF), was defined as a keratinocyte product induced by contact allergens comparatively late in the induction phase of ACD.[22] This rather late induction peaking at 12 h following application of a contact allergen to skin and the downmodulatory effect that IL-10 exerts on APC such as macrophages or dendritic cells suggested a counterregulatory role for IL-10 in ACD. In fact, it was demonstrated that IL-10 inhibits the accessory functions of epidermal LC and converts these cells from potent inducers of an immune response to tolerogenic APC.[23] In agreement with that later experiments showed that injection of IL-10 into murine skin prior to application of a contact allergen resulted in the induction of allergen-specific tolerance.[24] This was further corroborated by other experiments demonstrating that it was impossible to transfer CHS by T cells from animal to animal during the peak phase of IL-10 production in regional lymph nodes.[25] Neutralizing mAb to IL-10, however, completely restored the sensitizing capacity of allergen-treated T cells to transfer sensitization upon naïve mice. These findings suggested a counterregulatory role for IL-10 in the induction and elicitation phase of ACD.

In support of that studies using mAb-injections during the elicitation phase of CHS in mice showed enhanced ear swelling reactions in animals injected with neutralizing anti-IL-10 mAb. Also, experiments using IL-10-deficient mice showed exaggerated inflammatory reactions to contact allergens in mice lacking IL-10 in comparison to wild-type littermates.[26] IL-10-deficient mice also showed reduced influx of inflammatory cells and edema following application of the irritant croton oil.

Further studies in IL-10-deficient mice demonstrated an effect of IL-10 on the migration of LC following application of contact allergens.[27] It was shown that IL-10-deficient mice show enhanced LC migration. This effect was probably mediated by an enhanced production of IL-1 and TNF-α in the skin of mutant mice, as injection of anti-TNF-mAb or IL-1R-antagonist lead to an enhanced reduction of LC in the epidermis of mutant mice following application of a contact sensitizer.

In a low-zone tolerance system the counterregulatory role of IL-10 in ACD was further supported.[28] In this system, the application of low, nonsensitizing doses of allergens leads to the induction of allergen-specific tolerance. Tolerance induction is mediated by IL-10-producing $CD8^+$ T cells. Neutralization of IL-10 during low-zone application of haptens or use of IL-10-deficient mice abrogates tolerance induction.

In aggregate, the data support a role of IL-10 as an important counterregulatory agent in ACD reactions.

Interleukin-4

There has been controversial discussion on the role of IL-4 on ACD reactions. Initial studies using IL-4-deficient mice and the contact sensitizer OX did not observe any difference in IL-4-deficient mice to mount an ACD reaction as compared to their wild-type littermates.[25] In contrast, Asada et al reported on the production of IL-4 during the elicitation phase of ACD as an important counterregulatory factor, as neutralizing IL-4 mAb applied prior to hapten challenge resulted in enhanced ear swelling reactions.[29] In marked difference, studies published by Salerno et al report on an inhibition of the manifestation of ACD when anti-IL-4 mAb were applied to mice during the elication phase.[30]

In a series of elegant studies, Traidl et al demonstrated that the importance of IL-4 for the process of sensitization varies depending on the allergen chosen.[31] Using IL-4-deficient animals these investigators showed that transgenic animals showed a markedly reduced magnitude and duration of ACD reactions following application of the hapten DNFB compared to wild-type mice. This attenuation was accompanied by a significant reduction of edema and cellular infiltrates in the dermis and a lacking induction of IL-10 mRNA expression in skin. Furthermore, in adoptive transfer experiments, mice failed to exhibit ACD after injection of lymph node cells obtained from sensitized IL-4-deficient mice. In contrast, CHS reactions of transgenic mice to the hapte OX were indistinguishable from wild type controls and completely normal.

In aggregate, the data support an important proinflammatory role for IL-4 in the induction of ACD responses to certain haptens.

Conclusion

This chapter has summarized the role of cytokines mainly during the induction phase of ACD. It was shown that certain cytokines such as IL-1β or TNF-α are essential for the induction of ACD reactions by mediating early proinflammatory processes. Other cytokines such as IL-12 and IL-18 are more involved in directing the immune response towards a Tc1-mediated reaction. IL-10, on the other hand seems to have counterregulatory functions by limiting the extent of inflammation and also exerting tolerizing functions on local APC. IL-4 might be involved in directing the immune response towards a more Th2-like response during the elicitation phase of ACD, eventually to limit the extent of the immune reaction and protecting the body from harmful side effects.[32]

In summary, cytokines are crucial for the successful initiation, propagation and termination of ACD reactions and therefore also offer potential tools and targets for therapeutic applications.

References

1. Enk AH, Katz SI. Early molecular events in the induction phase of contact sensitivity. Proc Natl Acad Sci 1992; 89:1398-1402.
2. Enk AH. Allergic contact dermatitis: Understanding the immune response and potential for targeted therapy using cytokines. Molecular Med Today 1997; 3:423-428.
3. Enk AH, Angeloni VL, Udey MC et al. An essential role for Langerhans cell-derived IL-1β in the initiation of primary immune responses in the skin. J Immunol 1993; 150:3698-3704.
4. Enk AH, Katz SI. Contact sensitivity as a model for T-cell activation in skin. 1995; 105:80S-83S.
5. Shornick LP, De Togni P, Mariathasan S et al. Mice deficient in IL-1β manifest impaired contact hypersensitivity to trinitrochlorobenzene. J Exp Med 1996; 183:1427-1436.
6. Zheng H, Fletcher D, Kozak W et al. Resistance to fever induction and impaired acute-phase response in interleukin-1β deficient mice. Immunity 1995; 3:9-19.
7. Nakae S, Naruse-Nakajima CH, Sudo K et al. IL-1α, but not IL-1β, is required for contact-allergen-specific T cell activation during the sensitisation phase in contact hypersensitivity. Internat Immunology 2001; 13:1471-1478.
8. Piguet PF, Grau GE, Hauser C et al. Tumor necrosis factor is a critical mediator in hapten induced irritant and contact hypersensitivity reactions. J Exp Med 1991; 173:673.
9. Schwarzenberger K, Udey MC. Contact allergens and epidermal proinflammatory cytokines modulate Langerhans cell E-cadherin expression in situ. J Invest Dermatol 1996; 106:553-558.
10. Immune and inflammatory responses in TNFα-deficient mice: A critical requirement for TNFα in the formation of primary B cell follicles, follicular dendritic cell networks and germinal centers and in the maturation of the humoral immune response. J Exp Med 1996; 184:1397-1411.
11. Wang B, Fujisawa H, Zhuang L et al. Depressed langerhans cell migration and reduced contact hypersensitivity response in mice lacking TNF receptor p75. J Immunology 1997; 159:6148-6155.
12. Kondo S, Wang B, Fujisawa H et al. Effect of gene-targeted mutation in TNF receptor (p55) on contact hypersensitivity and ultraviolet B-induced immunosuppression. J Immunology 1995; 155:3801-3805.
13. Biedermann T, Kneilling M, Mailhammer R et al. Mast cell control neutrophil recruitment during T cell-mediated delayed-type hypersensitivity reactions through tumor necrosis factor and macrophage inflammatory protein 2. J Exp Med 2000; 192:1441-1451.
14. Müller G, Saloga J, Germann T et al. Identification and induction of human keratinocyte-derived IL-12. J Clin Invest 1994; 94:1799.

15. Müller G, Saloga J, Germann T et al. IL-12 as mediator and adjuvant for the induction of contact sensitivity in vivo. J Immunology 1995; 155:4661-4668.
16. Riemann H, Schwarz A, Grabbe St et al. Neutralization of IL-12 in vivo prevents induction of contact hypersensitivity and induces Hapten-specific tolerance. J Immunology 1996; 156:1799-1803.
17. Ushio H, Tsuji RF, Szczepanik M et al. IL-12 reverses established antigen-specific tolerance of contact sensitivity by affecting costimulatory molecules B7-1 (CD80) and B7-2 (CD86). J Immunology 1998; 160:2080-2088.
18. Gorbachev AV, DiIulio NA, Fairchild RL. IL-12 augments CD8⁺ T cell development for contact hypersensitivity responses and circumvents anti-CD154 antibody-mediated inhibition. J Immunology 2001; 167:156-162.
19. Schwarz A, Grabbe St, Mahnke K et al. Interleukin 12 break ultraviolet light induced immunosuppression by affecting CD8⁺ rather than CD4⁺ T cells. J Invest Dermatol 1998; 110:272-276.
20. Wang B, Feliciani C, Howell BG et al. Contribution of Langerhans cell-derived IL-18 to contact hypersensitivity. J Immunology 2002; 168:3303-3308.
21. Antonopoulos C, Cumberbatch M, Dearman RJ et al. Functional caspase-1 is required for Langerhans cell migration and optimal contact sensitisation in mice. J Immunology 2001; 166:3672-3677.
22. Enk AH, Katz SI. Identification and induction of keratinocyte-derived IL-10. J Immunology 1992; 149:92-95.
23. Enk AH, Angeloni VL, Udey MC et al. Inhibition of Langerhans cell antigen-presenting function by IL-10. J Immunology 1993; 151:2390.
24. Enk AH, Saloga J, Becker D et al. Induction of Hapten-specific tolerance by Interleukin 10 in vivo. J Exp Med 1994; 179:1397-1402.
25. Ferguson TA, Dube P, Griffith TS. Regulation of contact hypersensitivity by Interleukin 10. J Exp Med 1994; 179:1597-1604.
26. Berg DJ, Leach MW, Kuhn R et al. Interleukin 10 but not Interleukin 4 is a natural suppressant of cutaneous inflammatory responses. J Exp Med 1995; 182:99-108.
27. Wang B, Zhuang L, Fujisawa H et al. Enhanced epidermal Langerhans cell migration in IL-10 knockout mice. J Immunology 1999; 162:277-283.
28. Steinbrink K, Sorg C, Macher E. Low zone tolerance to contact allergens in mice: A functional role for CD8⁺ T helper type 2 cells. J Exp Med 1996; 183:759-768.
29. Asada H, Linton J, Katz SI. Cytokine gene expression during the elicitation phase of contact sensitivity: Regulation by endogenous IL-4. J Invest Dermatol 1997; 108:406-411.
30. Salerno A, Dieli F, Bella via A et al. Interleukin-4 is a critical cytokine in contact hypersensitivity. Immunology 1995; 84:404-409.
31. Traidl C, Jugert F, Krieg T et al. Inhibition of allergic contact dermatitis to DNCB but not to oxazolone in Interleukin-4-deficient mice. J Invest Dermatol 1999; 112:476-482.
32. Kitagaki H, Ono N, Hayakawa K et al. Repeated elicitation of contact hypersensitivity induces a shift in cutaneous cytokine milieu from a T helper cell type 1 to a T helper cell type 2 profile. J Immunology 1997; 159:2484-2491.

CHAPTER 8

The Use of Gene-Targeted Mice in Contact Hypersensitivity Research

Binghe Wang, Brandon G. Howell, Adam J. Mamelak and Daniel N. Sauder

Summary

Contact hypersensitivity (CHS), clinically presenting as allergic contact dermatitis, is a T cell-mediated cutaneous immune inflammatory response to reactive haptens. Although CHS is generally thought of as a Langerhans cell-dependent, Th1 type immune reaction, much is unknown about the complicated molecular mechanisms underlying CHS. With the advent of gene-targeting technology, a variety of gene knockout mouse models have been used in CHS research. Such studies have significantly contributed to our understanding of the molecular mechanisms comprising the CHS response. This review article summarizes about 70 different gene knockouts used to date in the field of CHS and discusses the contributions these studies have made toward understanding the molecular requirements for CHS initiation.

Introduction

Mutations in the genetic code have provided great insights into the function of genes. By interrupting or altering the final gene products, mutations have allowed investigators to directly elucidate the properties and roles of genes in vivo. For many years, the mutations obtained were limited, occurring only randomly during mouse breeding. Although certain strains of mice bearing these spontaneous mutations were utilized in research, the specific genetic defects in most of these cases were never fully characterized.

The ability to introduce genetic modifications into the germ line of complex organisms has been a long-standing goal of many investigators. And recently, this has been made possible via two technological breakthroughs: (a) the culturing of multipotent embryonic stem (ES) cells from mouse embryos, and (b) the introduction of designed mutations in mammalian cells by homologous DNA recombination between an incoming DNA and its cognate chromosomal sequence. The final result of these advancements has been the production of mice bearing null mutations in a specific gene, otherwise termed gene-targeted knockout (KO) mice[1] (Fig. 1).

Since 1989, when the first KO mouse was generated by Thompson and colleagues, hundreds of KOs have been created.[2] These genetically manipulated animals have been used extensively for biomedical research. The results of these trials have both challenged and refuted numerous biological paradigms. In the field of immunology for example, analysis of KOs has

Immune Mechanisms in Allergic Contact Dermatitis, edited by Andrea Cavani and Giampiero Girolomoni. ©2005 Eurekah.com.

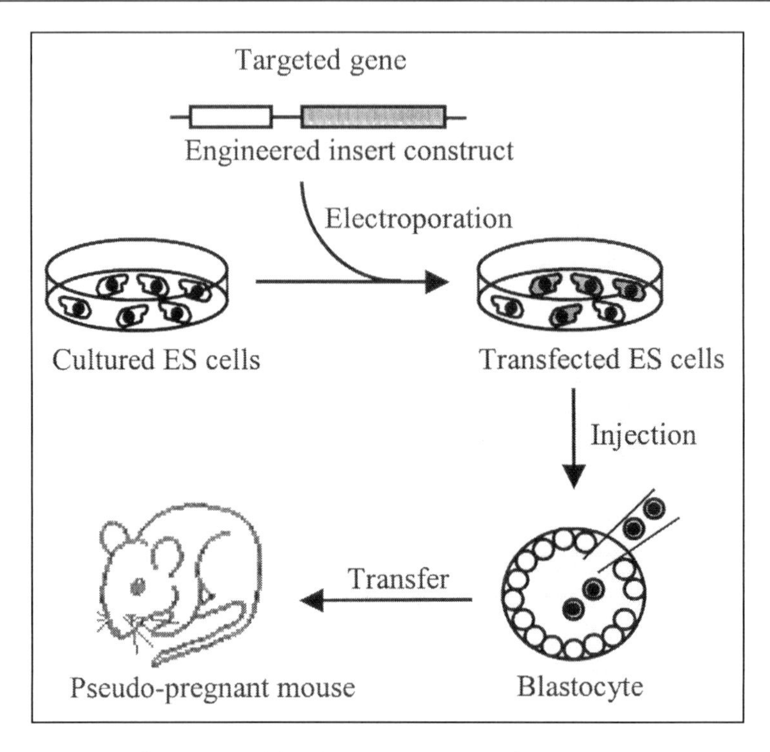

Figure 1. Generation of gene-targeted knockout mice. Engineered insertion construct containing the targeted gene is transfected into cultured embryonic stem (ES) cells through electroporation. The transfected ES cells are injected into a mouse blastocyte, and subsequently the embryo is implanted into a pseudo-pregnant female mouse.

clarified important intracellular signaling pathways and shed light on the regulatory mechanisms that govern immune responses.

Contact hypersensitivity (CHS), clinically presented as allergic contact dermatitis, is a T cell-mediated cutaneous immune response to reactive haptens. In general, CHS is thought of as a Langerhans cell (LC)-dependent, Th1 cell-mediated immune reaction. However, much is unknown about the complicated molecular mechanisms comprising the CHS response. During the past several years, a variety of KO mouse models have been used to assign biological functions to particular molecules in the CHS response.[3] Such studies have contributed significantly to our understanding of the mechanisms involved in CHS.

This chapter provides a summary of the KOs used to date for investigating molecular mechanisms underlying CHS. Also contained within are discussions of the new insights and molecular concepts gained from these studies. Each of the KOs discussed has been listed in (Tables 1-4) for convenience.

Gene-Targeted KO Mice with Abolished CHS Responses (Table 1)

When the skin is exposed to allergens, epidermal Langerhans cells (LCs) take up and process antigens, and migrate to regional lymph nodes (LNs). It is within the LNs where LCs present the antigenic peptides to naïve T cells. Antigen presentation and T cell activation require at least two signals. The interaction of T cell receptor (TCR) $\alpha\beta$ heterodimers with the MHC-peptide complex delivers the first signal. The second signal is derived from the interaction of LC costimulatory molecules with their receptors on T cells.

Table 1. KO mice with abolished CHS responses

Gene KO	Mouse Strain	Hapten	Ref.
TCRα–/–	129/J	PCI	4
MHC class I-/-	C57BL/6	DNFB	5
MHC class I/II-/-	C57BL/6	DNFB	5
LTα I-/-	BALB/c	OX, DNFB	6
LTβI-/-(w/o LN)	C57BL/6	OX,FITC	6
CCR7-/-	BALB/c x 129/Sv	FITC	7
CD11α (LFA-1α)-/-	C57B1/6	DNFB	8
CD18 (LFA-1β)-/-	129 X C57BL/6	OX,TNCB	9
Fuc-TIV/Fuc-TVII-/-	C57BL/6J	DNFB	10
perforin-/-/gld	C57BL/6	DNFB	12

PCI: picryl chloride; DNFB: dinitrofluorobenzene; OX: oxazolone; FITC: fluorescein isothiocyanate; TNCB: trinitrochlorobenzene

The crucial role of signal 1 in CHS initiation has been demonstrated via studies on TCRα chain KO (deficient in αβ T cells), MHC class I KO and MHC class I/II double KO mice. The observation that CHS responses were completely abolished in TCRα KO mice confirms αβ T cells function as the critical effector cells in CHS.[4] CHS responses were also abolished in MHC class I KO and MHC class I/II double-KO mice, supporting a crucial role for the interaction between TCRαβ and MHC during CHS initiation.[5]

LNs are important secondary lymphoid organs for the CHS response since they provide an environment for LCs to present antigens to naive T cells. Lymphotoxin (LT) α KO mice are essentially devoid of all LNs. These animals could not mount a CHS response.[6] LTβ KO mice have a similar phenotype, however 75% of these animals retain a subset of mucosa-associated LNs. Interestingly, these animals demonstrated a normal CHS response. Furthermore, the 25% of LTβ KO mice completely lacking LNs failed to mount a CHS response.[6] These results suggested that the ability of LTβ KO mice to generate a CHS response depends on the presence of LNs.

Chemokines are a group of chemotactic cytokines that bind specific receptors expressed on target cells, and they play an essential role in the migration of LCs from the skin to draining LNs. The CC chemokine receptor CCR7 has been shown to be important in CHS initiation. Resting epidermal LCs do not express detectable levels of CCR7. During LC migration from the epidermis to dermis, CCR7 is up-regulated. Lymphatic endothelial cells produce CCL21/ Secondary Lymphoid Tissue Chemokine (SLC), a cognate ligand of CCR7, and attract LCs into the lymphatic vessels. CCR7 is also important for the final positioning of LCs in the LNs, because CCL21/SLC is expressed by high endothelial venules (HEV) and stromal cells in the T cell areas of LNs. As well, CCL19/EBV-induced molecule 1 ligand chemokine (ELC), another ligand for CCR7, is produced by the resident dendritic cells (DCs) in the T cell areas. As anticipated, deletion of CCR7 led to complete loss of CHS responses in mice.[7]

Adhesion molecules control cell-cell attachment and cell transendothelial migration. The integrin subfamily is involved in both cell-matrix and cell-cell interactions. KO studies have demonstrated the importance of the β2-integrin LFA-1 (lymphocyte function-associated antigen-1) in CHS. In both CD11a KO mice (deficient in the LFA-1 α chain) and CD18 KO mice (deficient in the LFA-1 β chain), CHS responses were abolished.[8,9] Other adhesion molecule subfamilies are also involved in the CHS response. The α-1,3-fucosyltransferases (Fuc-Ts) are thought to regulate expression of selectin receptors. Two of the six Fuc-T loci, Fuc-TIV and

Fuc-TVII, control leucocyte and HEV expression of selectin ligands. Importantly, a recent study has demonstrated complete loss of the CHS response in Fuc-TIV/TVII double KO mice.[10]

Cytotoxic T lymphocytes (CTL) mediate cell killing through two pathways: the perforin pathway and Fas/FasL pathway. Perforin KO mice, spontaneous mutant Fas-deficient *lpr* mice, and Fas ligand (FasL)-deficient *gld* mice all had normal CHS responses. However, perforin KO/*gld* mice (deficient in both perforin and FasL) did not show any CHS response.[12] This observation suggests that CHS requires simultaneous activation of these cytotoxic pathways.

Gene-Targeted KO Mice with Decreased CHS Responses (Table 2)

The CD4 and CD8 molecules are two coreceptors on T cells that stabilize and increase the avidity of interaction between the TCR and peptide-MHC determinants on antigen-presenting cells. $CD4^+$ T cells recognize peptides in the antigen-binding pocket of MHC class II molecules, whereas $CD8^+$ T cells recognize peptides in the antigen-binding pocket of MHC class I. Traditionally, it is believed that CHS represents a prototypical delayed-type hypersensitivity (DTH) response, mediated by $CD4^+$ T cells and down-regulated by $CD8^+$ T cells.

The effector role of $CD4^+$ T cells in CHS has been supported by studies using CD4 KO mice and CD83 KO mice. CD4 KO mice (deficient in $CD4^+$ T cells) had a decreased CHS response.[13] CD83 is a cell surface molecule expressed by thymic epithelial cells and DCs. The engagement of CD83 with precursor cells is required for the generation of $CD4^+$ T cells. CD83 KO mice had a selective 75-90% reduction in peripheral $CD4^+$ T cells and consequently an impaired CHS response.[14] Some studies however, have suggested the contrary. For example, CHS responses were abolished in MHC class I KO mice but exaggerated in MHC class II KO mice.[5] These observations support the effector role of $CD8^+$ T cells, not $CD4^+$ T cells, in mediating the CHS response.

Recently, a study from our laboratory has clarified these conflicting data. With the help of both CD4 KO mice and CD8 KO mice we have demonstrated that both $CD4^+$ Th1 and $CD8^+$ Tc1 cells contribute to the development of CHS. In this study, both CD4 KO mice and CD8 KO mice exhibited a decreased CHS response.[15] As well, in vivo depletion of either $CD8^+$ T cells from CD4 KO mice or $CD4^+$ T cells from CD8 KO mice virtually abolished CHS responses.

The important role of the costimulatory signals, released by the interaction between costimulatory molecules and their receptors in CHS, has also been confirmed by KO studies. The contributions of the B7-CD28 interaction were verified when B7-2 KO, B7-1/-2 double KO, and CD28 KO mice all displayed decreased CHS responses.[16,17] An additional costimulatory pathway known as the Ox40/Ox40L pathway may be also involved in the initiation of CHS. This comes from the observation that Ox40L KO mice showed impaired CHS responses.[18] Further studies have demonstrated a fundamental role for the interaction between CD40 and CD40L in CHS. CD40L KO mice had defective DC migration associated with a decreased CHS response.[19]

The selectin subfamily of adhesion molecules mediates adhesion of leukocytes to endothelial cells. A decreased CHS response has been reported in L-selectin KO,[20,21] P-selectin KO,[22] and E-/P-selectin double KO,[23] as well as PSGL-1 (p-selectin glycoprotein ligand 1) KO mice.[24] The aforementioned Fuc-TVII molecule plays an essential role in controlling E- and P-selectin ligand expression in Th1 and Tc1 lymphocytes. As with the selectin KO mice, Fuc-TVII KO mice demonstrated a substantial CHS impairment.[10,11] In addition, α1-integrin KO,[25] ICAM-1 KO,[26,27] and L-selectin/ICAM-1 double KO mice[28] have all demonstrated a reduced CHS response. This directly illustrates the importance of selectins to the pathogenesis of CHS.

Table 2. KO mice with decreased CHS responses

Gene KO	Mouse Strain	Hapten	Ref.
CD4-/-	C57BL/6	DNFB	13
CD83-/-	129Sv x C57BL/6	OX	14
CD8-/-	C57BL/6	DNFB, OX	15
B7-2-/-	129SvJ	OX	16
B7-1/-2-/-	OX	OX	16
CD28-/-	057BL/6	DNFB, OX	17
Ox4OL-/-	BALB/c	DNFB, OX, FITC	18
CD4OL-/-	C57BL/6	FITC	19
L-selectin	129Sv x C57BL/6	OX	20,21
P-selectin-/-	129Sv x C57BL/6	OX	22
E-/P-selectin-/-	129SvJ x C57BL/6	OX	23
PSGL-1-/-	129SvJ x C57BL/6	OX	24
α1 integrin-/-	BALB/c	FITC	*25*
ICAM-1-/-	129Sv x C57BL/6J	DNFB	26,27
L-selectin/ICAM-1-/-	C57BL/6	OX	28
Fuc-TVII-/-	C57BL/6J	DNFB, FITC	10,11
IFN-γR-/-	129Sv/Ev	OX, TNCB	29
IFN-γR2-/-	129Sv x C57BL/6	FITC	30
IL-1 α-/-	C57BL/6J	TNCB	31
IL-1 β-/-	129Sv x 057BL/6J	TNCB	32,33
IL-1 αβ-/-	C57BL/6J	TNCB	31
IL-3-/-	BALB/c	DNFB	34
IL-4-/-	BALB/c	DNFB	35
IL-6-/-	C57BL/6	OX	36
GM-CSF/IL-3-/-	C57BL/6, BALB/c	OX, FITC	37
TNF-α-/-	129Sv x C57BL/6	OX	38
TNFR2-/-	C57BL/6	OX	39
TNFR1/2-/-	129SvJ	DNFB	40
LTα-/-	BALB/c	FITC	8
STAT6-/-	C57BL/6	DNFB, TNCB, OX	41
Caspase-1-/-	129Sv x C57BL/6	DNCB, OX	42
MCP-1-/-	129Sv x C57BL/6	DNFB	43
IP-1O-/-	129Sv/J	DNFB	44
OPN-/-	C57BL/6	TNCB	45
C5aR-/-	C57BL/6 x 129	PCI	46
Igh-6-/-	C57BL/6	PCI	47
Thy-1-/-	129Sv x C57BL/6	TNCB,OX	48
MMP -3-/-	C57BL/10RIII	DNFB	49
Re1B-/-	129Sv x C57BL/6	DNFB, OX	50
FcRγ-/-	129Sv x C57BL/6	OX	51
GADD45γ-/-	129SvE x C57BL/6	DNFB	52
CD39-/-	129SvJ x C57BL/6	OX	53

Cytokines play effector and regulatory roles in CHS. It is generally believed that Th1 cytokines (i.e., interferon (IFN)-γ) play effector roles in CHS while Th2 cytokines (i.e., interleukin (IL)-4, IL-10 and IL-13) play a down-regulatory role. Certain KO studies have supported an effector role for IFN-γ in CHS. For example, IFN-γ receptor (IFN-γR) KO mice had diminished dermal mononuclear cell infiltrates and epidermal microabscesses in CHS reactions.[29]

IFN-γR2 KO mice had decreased ear swelling responses.[30] It is of importance however, that certain cytokine KO studies have suggested CHS is not strictly a Th1-mediated response. For example, IL-4 KO mice showed a diminished, rather than an enhanced CHS response to DNCB, implying an effector role for this Th2 cytokine in CHS.[35] Likewise, observations of a decreased CHS response in mice deficient in STAT6 (signal transducer and activator of transcription 6) further support this effector role of Th2 cytokines in CHS, because STAT6 plays a central role in the signaling pathways of IL-4 and IL-13.[41]

IL-1, IL-6, IL-18 and tumor necrosis factor (TNF)-α are all important proinflammatory cytokines involved in CHS responses. Decreased CHS responses have been seen in IL-1α KO,[31] IL-1β KO,[32,33] IL-6 KO,[36] TNF-α KO,[38] TNF receptor 2 (TNFR2) KO,[39] and TNFR1/2 double KO mice.[40] Also, an impaired CHS response was demonstrated in caspase-1 KO mice (deficient in mature IL-1β and IL-18).[42]

There have also been recent experiments to suggest involvement of IL-3 in the CHS response. IL-3 is a hematopoietic growth factor involved in the development, survival and function of mast cells and basophils. Curiously, hematopoiesis was not impaired in IL-3 KO mice, whereas CHS reactions were diminished.[34]

Chemokines further contribute to the pathogenesis of CHS. Monocyte chemoattractant protein-1 (MCP-1) is a CC chemokine that attracts monocytes, memory T cells and natural killer cells. MCP-1 KO mice demonstrated decreased inflammatory cell infiltrates in CHS reactions.[43] IFN-γ-inducible protein 10 (IP-10/CXCL10) is a chemoattractant for activated T cells. Expression of IP-10 is seen in Th1-type inflammatory diseases, where it is thought to play an important role in recruiting activated T cells into sites of tissue inflammation. Deletion of IP-10 resulted in an impaired CHS response.[44] Osteopontin (OPN) exerts dual functions both as a chemokine and as a component of ECM. In its function as a chemokine, OPN attracts monocytes to sites of inflammation and stimulates chemotactic endothelial cell migration. OPN also induces LC migration by interacting with CD44 and α5-integrin. OPN KO mice had a significantly reduced CHS response associated with an impaired ability to attract LCs to draining LNs.[45]

KO studies have supported the implication of the complement system in the development of CHS. For example, C5a receptor (C5aR) KO mice had an impaired CHS response.[46] Decreased CHS responses were also seen in the immunoglobulin heavy-chain-6 (Igh-6) KO mice, indicating the involvement of complement-fixing immunoglobulins in CHS.[47]

A range of other molecules may be involved in the CHS response as well. One example is Thy-1 (CD90); a cell surface molecule expressed on murine T cells. Thy-1 KO mice had a reduced CHS response, suggesting a role for Thy-1 in CHS.[48] Matrix metalloproteinases (MMPs) are a family of zinc-dependent endopeptidases that degrade ECM. MMPs play an important role in the migration of T cells and macrophages through the skin. MMP-3 KO mice showed a markedly reduced CHS response.[49] The *relB* gene encodes a transcription factor that belongs to the family of nuclear factor (NF)-κB/Rel proteins. RelB KO mice had an impaired CHS response, suggesting this family of transcription factors contribute to CHS pathogenesis.[50] The γ subunit of the Fc receptor for Igs (FcRγ) is an essential component of the IgE high-affinity receptor (FcεRI) and the low-affinity receptor for IgG (FcγγRIII). Targeted disruptions of this FcR γ subunit resulted in a decreased CHS response.[51] GADD45γ, a member of the GADD45 family, mediates effector functions of Th1 cells. GADD45γ deficiency also caused a reduction in CHS.[52] CD39, the endothelial ecto-nucleoside triphosphate diphoshydrolase (NTPDase), regulates vascular inflammation and thrombosis by hydrolyzing ATP and ADP. LCs in CD39-/- mice express no detectable ecto-NTPDase activity. The contact hypersensitivity response was severely attenuated in CD39-/- mice, suggesting a role for CD39 in the activation of T cells by LCs.[53]

Table 3. KO mice with an enhanced CHS response

Gene KO	Mouse Strain	Hapten	Ref.
TCRδ-/-	FVB	DNFB	54
MHC class II-/-	C57BL/6	DNFB	5
IL-10-/-	129Ola x C57BL/6	OX	55
	C57BL/6	FITC	56
TNFR1-/-	C57BL/6	DNFB	57
CCR5-/-	129Sv x ICR	FITC	58
CCR6	129SvJ x C57BL/6	DNFB	59
iNOS-/-	129SvEv x C57BL/6	DNFB	60
Ron TK-/-	129Ola x CD-1	DNFB	61
TSP-2-/-	129SvJ	OX	62
MMP9-/-	129Sv	DNFB	49
tenascin-c-/-	BALB/c	DNFB	63
NEP-/-	C57BL/6	DNFB	64
PKR-/-	129SvEv	DNFB	65

Gene-Targeted KO Mice with an Enhanced CHS Response (Table 3)

Enhanced CHS responses have been reported in certain KO mice, indicating a negative regulatory role for certain gene products. In contrast to the effector role played by αβ T cells, γδ T cells may play a down-regulatory role in CHS. This idea arose from studies indicating that CHS responses were enhanced in TCRδ KO mice (deficient in γδ T cells).[54]

IL-10 is an important regulatory cytokine. The down-regulatory role of IL-10 in CHS has been confirmed by observations that IL-10 KO mice had an enhanced CHS response. In these mutant mice the CHS response was increased in both magnitude and duration as compared to wild-type mice.[55,56]

The CC chemokine receptors CCR5 and CCR6 are expressed on immature DCs and T lymphocytes. These receptors bind a range of inflammatory chemokines. An enhanced CHS response has been reported in CCR5 KO[58] and CCR6 KO mice.[59] These findings indicate a novel role for CCR5 and CCR6 in down-modulating T cell-dependent immune responses.

Ron receptor, a member of the tyrosine kinase (TK) receptor family, is involved in many aspects of the immune response including antigen presentation, macrophage activation, and nitric oxide (NO) regulation. KO mice with targeted deletion of the Ron TK domain had an enhanced CHS response, suggesting Ron-mediated cytoplasmic signaling may be required for the limitation of inflammatory responses.[61]

Angiogenesis and enhanced microvascular permeability are hallmarks of a variety of inflammatory diseases. Thrombospondin-2 (TSP-2) is an endogenous angiogenesis inhibitor whose expression has shown to be up-regulated in inflamed skin. Deficiency in TSP-2 significantly enhanced and prolonged the CHS response.[62] It is thought that the antiangiogenic activity of TSP-2 exerts this effect via limiting the extent and duration of edema formation and inflammatory cell infiltration throughout the CHS reaction.

Tenascin-C is an ECM glycoprotein and KO mice deficient in this molecule mount an enhanced CHS response. This suggests a negative regulatory role for tenascin-C in cutaneous inflammation.[63] Sensory nerve-derived neuropeptides such as substance P are also found in the ECM and demonstrate a number of proinflammatory bioactivities. The cell surface metalloprotease termed neutral endopeptidase (NEP) is the principal proteolytic substance P-degrading enzyme. Enhanced CHS responses were demonstrated in NEP KO mice, indicating a significant role for cutaneous neuropeptides in the CHS reaction.[64]

Table 4. KO mice with normal CHS responses

Gene KO	Mouse Strain	Hapten	Ref.
TCR δ-/-	C57BL/6	DNFB,PCI	54
B7 -1-/-	129SvJ	OX	15
E-selectin-/-	129Sv x C57BL/6	OX	22
PECAM-1-/-	C57BL/6	OX	66
CD 40 L-/-	C57BL/6	DNFB	67
IFN-γ-/-	C57BL/6	DNFB	68
IL-1 β-/-	129Sv x C57BL/6J	OX	69
IL-2-/-	129Ola x C57BL/6	OX	36
IL-4-/-	BALB/c	OX	70
IL-12 P^{40}-/-	BALB/c	DNFB	71
LTβ-/- (w/LN)	C57BL/6	OX,FITC	8
GM-CSF-/-	C57BL/6, BALB/c	OX,FITC	37
Perforin-/-	C57BL/6	DNFB	12
Fuc-TIV-/-	C57BL/6J	DNFB	10
Tenascin-C-/-	C3H/HeN, C57BL/6	DNFB	63
LTA_4hydrolaes-/-	129Ola x 129SvEv	DNFB	72
IRAK-/-	129Sv x C57BL/6	DNFB	73
XPA-/-	C57BL/6	PCI	74
CSB-/-	C57BL/6	PCI	74
XPC-/-	C57BL/6	PCI	75
TTD-/-	C57BL/6	PCI	76

The IFN-induced and dsRNA-activated kinase PRK is a well-characterized component of IFN-regulated antiviral and antiproliferative responses. PKR may also play a role in the regulation of immune responses since PKR KO mice demonstrated an enhanced CHS response. This suggests a potential suppressive role for this kinase in CHS.[65]

Keratinocytes and LCs are both capable of NO synthesis. This process is mediated by NO synthases, including the inducible isoform of NO synthase (iNOS). NO is clearly involved in cutaneous immune and inflammatory reactions. However, an unexpected observation came from iNOS KO mice which demonstrated an enhanced, rather than impaired, CHS response.[60] The reasons for this are yet to be determined.

Gene-Targeted KO Mice with Normal CHS Responses (Table 4)

Although numerous KO studies have yielded predictable results, investigations with KOs have sometimes provided unanticipated data. Certain molecules are believed to be crucial to the CHS response. Quite unexpectedly, KO studies have demonstrated that these molecules could safely be deleted from mice without significantly affecting CHS. For instance, a normal CHS response has been found in IFN-γ KO,[68] IL-2 KO,[36] and IL-12 p40 KO mice.[71] Thus, it appears that various molecular combinations have overlapping effects in CHS. CHS is likely a redundant system utilizing several molecules with shared function. That is, in the event that certain important molecules are absent or deleted from the CHS reaction, other molecules of similar function may take their place.

Normal CHS responses have been observed in KO mice with targeted deletion of established cutaneous inflammatory mediators. For example, leukotriene (LT) B_4 is a potent inflammatory mediator, however LTA_4 hydrolase KO mice (deficient in LTB_4) had a normal CHS response.[72] IL-1 receptor-associated kinase (IRAK) is involved in IL-1 and IL-18 signal transduction. IRAK KO mice however had a normal CHS response, suggesting involvement of an IRAK-independent pathway in CHS pathogenesis.[73]

Other notable KO mice possessing a normal CHS response include B7-1 KO,[15] E-selectin KO,[22] PECAM-1 (platelet/endothelial cell adhesion molecule-1) KO,[66] and GM-CSF KO mice.[37]

Exposure to ultraviolet (UV)B radiation impairs Th1 immune responses. This immunosuppression is instigated in part by the DNA-damaging effects of UVB. Ordinarily, UVB-induced DNA damage is repaired by a process known as nucleotide excision repair (NER). Since enhanced NER can strongly counteract immunosuppression, CHS experiments were pursued using four strains of NER-deficient mice with targeted disruption of XPA, CSB, XPC, and TTD. Unexpectedly, all these mutant mice had a normal CHS response.[74-76] This suggests that molecules of redundant function in the NER pathway may compensate for the deficiency in DNA repair in these mice, or simply that CHS responses are unaffected by NER status.

TCRδ KO mice with a C57BL/6 background demonstrated a normal CHS response, however the mutants with a FVB background had an enhanced CHS response.[54] Similarly, a normal CHS response was seen in tenascin-C KO mice with C3H/HeN and C57BL/6N backgrounds, but the KO mice with BALB/c backgrounds had an enhanced CHS response.[63] This illustrates that CHS responses in KO mice can vary with genetic background.

In addition to the murine genetic background, different hapten types may also result in qualitatively different immune responses. For example, IL-1β KO mice had a normal CHS response to OX, but a decreased response to trinitrochlorobenzene (TNCB).[69] IL-4 KO mice demonstrated a normal CHS response to OX, but a decreased response to dinitrofluorobenzene (DNFB).[70] Furthermore, CD40L KO mice showed a normal CHS response to DNFB, but a decreased CHS response to FITC.[67]

Concluding Remarks

Gene-targeted KO technology now permits researchers to generate precise and long-standing alterations of a particular gene, so that its functions can be deduced in vivo. The analysis of such mutants has significantly increased our understanding of the roles of particular molecules in the immune system. CHS research has certainly been revolutionized by these studies. Interestingly, many of these CHS studies have yielded unpredicted findings, and challenged the immunological paradigms provided by previous work.

We now propose that CHS is a LC-dependent, αβ T cell-mediated, heterogeneous skin immune response to reactive haptens. As well, CHS differs from classic DTH responses. DTH is mediated by $CD4^+$ Th1 cells, whereas both $CD4^+$ Th1 cells and $CD8^+$ Tc1 cells can function as effector cells in CHS. Furthermore, various molecules are involved in the development of CHS including costimulatory molecules, adhesion molecules, cytokines and chemokines. The CHS response is not necessarily mediated by Th1 cytokines, as Th2 patterns may also contribute depending on hapten used. Moreover, MMPs, neuropeptides, and other molecules may also regulate the development of CHS.

The immense impact that KO mice have had on CHS research is not only due to the fact that they help answer important questions, but also because they raise many more. For example, the discrepancies between CHS responses in IFN-γ KO and IFN-γ R KO mice have questioned the role of IFN-γ as the main effector cytokine in this immune reaction. As well, IL-3 is a growth factor for hematopoietic cells, yet the IL-3 KO mouse experiences alterations in the CHS response. How does IL-3 play a role in CHS? It is the desire of many investigators that these and other questions be answered by further studies in this field.

Recently, more-refined genetic techniques including tissue-specific and inducible gene KO mice have been developed. These techniques were developed as necessary measures to rescue mutant animals from inescapable developmental defects or death. These techniques are now being employed to directly address the issues of molecular compensation and redundancy in signaling pathways, and to determine the exact roles of particular molecules in the immune response. Indeed, we and other investigators look forward to the promises that the future of CHS research holds.

References

1. Koller BH, Smithies O. Altering genes in animals by gene targeting. Annu Rev Immunol 1992; 10:705-730.
2. Mak TW, Penninger JM, Ohashi PS. Knockout mice: A paradigm shift in modern immunology. Nature Rev Immunol 2001; 1(1):11-19.
3. Wang B, Feliciani C, Freed I et al. Insights into molecular mechanisms of contact hypersensitivity gained from gene knockout studies. J Leukoc Biol 2001; 70(2):185-191.
4. Szczepanik M, Anderson LR, Ushio H et al. γδ T cells from tolerized αβ T cell receptor (TCR)-deficient mice inhibit contact sensitivity-effector T cells in vivo, and their interferon-gamma production in vitro. J Exp Med 1996; 184(6):2129-2139.
5. Bour H, Peyron E, Gaucherand M et al. Major histocompatibility complex class I-restricted $CD8^+$ T cells and class II-restricted $CD4^+$ T cells, respectively, mediate and regulate contact sensitivity to dinitrofluorobenzene. Eur J Immunol 1995; 25(11):3006-3010.
6. Rennert PD, Hochman PS, Flavell RA et al. Essential role of lymph nodes in contact hypersensitivity revealed in lymphotoxin-α-deficient mice. J Exp Med 2001; 193(11):1227-1238.
7. Forster R, Schubel A, Breifeld D et al. CCR7 coordinates the primary immune response by establishing functional microenvironments in secondary lymphoid organs. Cell 1999; 99(1):23-33.
8. Schmits R, Kundig TM, Baker DM et al. LFA-1-deficient mice show normal CTL responses to virus but fail to reject immunogenic tumor. J Exp Med 1996; 183(4):1415-1426.
9. Grabbe S, Varga G, Beissert S et al. β_2 integrins are required for skin homing of primed T cells but not for priming naïve T cells. J Clin Invest 2002; 109(2):183-192.
10. Smithson G, Rogers CE, Smith PL et al. Fuc-TVII is required for T helper 1 and T cytotoxic 1 lymphocyte selectin ligand expression and recruitment in inflammation, and together with Fuc-TIV regulates naïve T cell trafficking to lymph nodes. J Exp Med 2001; 194(5):601-614.
11. Erdmann I, Scheidegger EP, Koch FK et al. Fucosyltransferase VII-deficient mice with defective E-, P-, and L-selectin ligands show impaired $CD4^+$ and $CD8^+$ T cell migration into the skin, but normal extravasation into visceral organs. J Immunol 2002; 168(5):2139-2146.
12. Kehren J, Desvignes C, Krasteva M et al. Cytotoxicity is mandatory for $CD8^+$ T cell-mediated contact hypersensitivity. J Exp Med 1999; 189(5):779-786.
13. Kondo S, Beissert S, Wang B et al. Hyporesponsivenmess in contact hypersensitivity and irritant contact dermatitis in CD4 gene targeted mouse. J Invest Dermatol 1996; 106(5):993-1000.
14. Fujimoto Y, Tu L, Miller NS et al. CD83 expression influences $CD4^+$ T cell development in the thymus. Cell 2002; 108(6):755-767.
15. Wang B, Fujisawa H, Zhuang L et al. $CD4^+$ Th1 and $CD8^+$ type 1 cytotoxic T cells both play a crucial role in the full development of contact hypersensitivity. J Immunol 2000; 165(12):6783-6790.
16. Rauschmayr-Kopp T, Williams IR, Borriello F et al. Distinct roles for B7 costimulation in contact hypersensitivity and humoral immune responses to epicutaneous antigen. Eur J Immunol 1998; 28(12):4221-4227.
17. Kondo S, Kooshesh F, Wang B et al. Contribution of the CD28 molecule to allergic and irritant-induced skin reactions in CD28 -/- mice. J Immunol 1996; 157(11):4822-4829.
18. Chen AI, McAdam AJ, Buhlmann JE et al. Ox40-ligand has a critical costimulatory role in dendritic cell: T cell interactions. Immunity 1999; 11(6):689-698.
19. Moodycliffe AM, Shreedhar V, Ullrich SE et al. CD40-CD40 ligand interactions in vivo regulate migration of antigen-bearing dendritic cells from the skin to draining lymph nodes. J Exp Med 2000; 191(11):2011-2020.
20. Xu J, Grewal IS, Geba GP et al. Impaired primary T cell responses in L-selectin-deficient mice. J Exp Med 1996; 183(2):589-598.
21. Tedder TF, Steeber DA, Pizcueta P. L-selectin-deficient mice have impaired leukocyte recruitment into inflammatory sites. J Exp Med 1995; 181(6):2259-2264.

22. Subramaniam M, Saffaripour S, Watson SR et al. Reduced recruitment of inflammatory cells in a contact hypersensitivity response in P-selectin-deficient mice. J Exp Med 1995; 181(6):2277-2282.
23. Staite ND, Justen JM, Sly LM et al. Inhibition of delayed-type contact hypersensitivity in mice deficient in both E-selectin and P-selectin. Blood 1996; 88(8):2973-2979.
24. Yang J, Hirata T, Croce K et al. Targeted gene disruption demonstrates that P-selectin glycoprotein ligand 1 (PSGL-1) is required for P-selectin-mediated but not E-selectin-mediated neutrophil rolling and migration. J Exp Med 1999; 190(12):1769-1782.
25. de Fougerolles AR, Aprague AG, Nickerson-Nutter CL et al. Regulation of inflammation by collagen-binding integrin $\alpha 1\beta 1$ and $\alpha 2\beta 2$ in models of hypersensitivity and arthritis. J Clin Invest 2000; 105(6):721-729.
26. Sligh Jr JE, Ballantyne CM, Rich SS et al. Inflammatory and immune responses are impaired in mice deficient in intercellular adhesion molecule 1. Proc Natl Acad Sci USA 1993; 90(19):8528-8533.
27. Xu H, Gonzalo JA, St Pierre Y et al. Leukocytosis and resistance to septic shock in intercellular adhesion molecule 1-deficient mice. J Exp Med 1994; 180(1):95-109.
28. Steeber DA, Tang MLK, Green NE et al. Leukocyte entry into sites of inflammation requires overlapping interactions between the L-selectin and ICAM-1 pathways. J Immunol 1999; 163(4):2176-2186.
29. Saulnier M, Huang S, Aguet M et al. Role of interferon-γ in contact hypersensitivity assessed in interferon-γ receptor-deficient mice. Toxicology 1995; 102(3):301-312.
30. Lu B, Ebensperger C, Dembic Z et al. Targeted disruption of the interferon-γ receptor 2 gene results in severe immune defects in mice. Proc Natl Acad Sci USA 1998; 95(14):8233-8238.
31. Nakae S, Naruse-Nakajima C, Sudo K et al. IL-1α, but not IL-1β, is required for contact-allergen-specific T cell activation during the sensitization phase in contact hypersensitivity. Int Immunol 2001; 13(12):1471-1478.
32. Shornick LP, De Togni P, Mariathasan S et al. Mice deficient in IL-1β manifest impaired contact hypersensitivity to trinitrochlorobenzene. J Exp Med 1996; 183(4):1427-1436.
33. Shornick LP, Bisarya AK, Chaplin DC et al. IL-1β is essential for Langerhans cell activation and antigen delivery to the lymph nodes during contact sensitization: Evidence for a dermal source of IL-1β. Cell Immunol 2001; 211(2):105-112.
34. Mach N, Lantz CS, Galli SJ et al. Involvement of interleukin-3 in delayed-type hypersensitivity. Blood 1998; 91(3):778-783.
35. Traidl C, Jugert F, Krieg T et al. Inhibition of allergic contact dermatitis to DNFB but not to oxazolone in interleukin-4-deficient mice. J Invest Dermatol 1999; 112(4):476-482.
36. Hope JC, Campbell F, Hopkins SJ. Deficiency of IL-2 or IL-6 reduces lymphocyte proliferation, but only IL-6 deficiency decreases the contact hypersensitivity response. Eur J Immunol 2000; 30(1):197-203.
37. Gillessen S, Mach N, Small C et al. Overlapping roles for granulocyte-macrophage colony-stimulating factor and interleukin-3 in eosinophil homeostasis and contact hypersensitivity. Blood 2001; 97(4):922-928.
38. Pasparakis M, Alexopoulou L, Episkopou V et al. Immune and inflammatory responses in TNFα-deficient mice: A critical requirement for TNFα in the formation of primary B cell follicular dendritic cell networks and germinal centers, and in the maturation of the humoral immune response. J Exp Med 1996; 184(4):1397-1441.
39. Wang B, Fujisawa H, Zhuang L et al. Depressed Langerhans cell migration and reduced contact hypersensitivity response in mice lacking TNF receptor p75. J Immunol 1997; 159(12):6148-6155.
40. Amerio P, Toto P, Feliciani C et al. Rethinking the role of tumour necrosis factor-α in ultraviolet (UV) B-induced immunosuppression: Altered immune response in UV-irradiated TNFR1R2 gene-targeted mutant mice. Br J Dermatol 2001; 144(5):952-957.
41. Yokozeki H, Ghoreishi M, Takagawa K et al. Signal transducer and activator of transcription 6 is essential in the induction of contact hypersensitivity. J Exp Med 2000; 191(6):995-1004.
42. Antonopoulos C, Cumberbatch M, Dearman RJ et al. Functional caspase-1 is required for Langerhans cell migration and optimal contact sensitization in mice. J Immunol 2001; 166(6):3672-3677.
43. Lu B, Rutledge BJ, Gu L et al. Abonormality in monocyte recruitment and cytokine expression in monocyte chemoattractant protein 1-deficient mice. J Exp Med 1998; 187(4):601-608.

44. Dufour JH, Dziejman M, Liu MT et al. IFN-γ-inducible protein 10 (IP-10; CXCL10)-deficient mice reveal a role for IP-10 in effector T cell generation and trafficking. J Immunol 2002; 168(7):3195-3204.
45. Weiss JM, Renkl AC, Maier CS et al. Osteopontin is involved in the initiation of cutaneous contact hypersensitivity by inducing Langerhans and dendritic cell migration to lymph nodes. J Exp Med 2001; 194(9):1219-1229.
46. Tsuji RE, Kawikova I, Ramabhadran R et al. Early local generation of C5a initiates the elicitation of contact sensitivity by leading to early T cell recruitment. J Immunol 2000; 165(3):1588-1598.
47. Tsuji RF, Geba GP, Wang Y et al. Required early complement activation in contact sensitivity with generation of local C5-dependent chemotactic activity, and late T cell interferon γ: A possible initiating role of B cells. J Exp Med 1997; 186(7):1015-1026.
48. Beissert S, He HT, Hueber A et al. Impaired cutaneous immune responses in Thy-1-deficient mice. J Immunol 1998; 161(10):5296-5302.
49. Wang M, Qin X, Mudgett JS et al. Matrix metalloproteinase deficiencies affect contact hypersensitivity: Stromelysin-1 deficiency prevents the response and gelatinase B deficiency prolongs the response. Proc Natl Acad Sci USA 1999; 96(12):6885-6889.
50. Weih F, Carrasco D, Durham SK et al. Multiorgan inflammation and hematopoietic abnormalities in mice with a targeted disruption of RelB, a member of the NF-κB/Rel family. Cell 1995; 80(2):331-340.
51. Zhang L, Tinkle SS. Chemical activation of innate and specific immunity in contact dermatitis. J Invest Dermatol 2000; 115(2):168-176.
52. Lu B, Yu H, Chow CW et al. GADD45γ mediates the activation of the p38 and JNK MAP kinase pathways and cytokine production in effector T_H1 cells. Immunity 2001; 14(5):583-590.
53. Mizumoto N, Kumamoto T, Robson S et al. CD39 is the dominant Langerhans cell-associated ecto-NTPDase: Modulatory roles in inflammation and immune responsiveness. Nature Med 8(3):358-365.
54. Girardi M, Glusac E, Ferrari C et al. Augmented allergic and irritant dermatitis in $\gamma\delta^+$ T-cell deficient mice is dependent on genetic background. J Invest Dermatol 2000; 114(4):762.
55. Berg DJ, Leach MW, Kuhn R et al. Interleukin 10 but not interleukin 4 is a natural suppressant of cutaneous inflammatory responses. J Exp Med 1995; 182(1):99-108.
56. Wang B, Zhuang L, Fujisawa H et al. Enhanced epidermal Langerhans cell migration in IL-10 knockout mice. J Immunol 1999; 162(1):277-283.
57. Kondo S, Wang B, Fujisawa H et al. Effect of gene-targeted mutation in TNF receptor (p55) on contact hypersensitivity and ultraviolet B-induced immunosuppression. J Immunol 1995; 155(6):3801-3805.
58. Zhou Y, Kurihara T, Ryseck RP et al. Impaired macrophage function and enhanced T cell-dependent immune response in mice lacking CCR5, the mice homologue of the major HIV-1 coreceptor. J Immunol 1998; 160(8):4018-4025.
59. Varona R, Villares R, Carramolino L et al. CCR6-deficient mice have impaired leukocyte hemeostasis and hypersensitivity responses. J Clin Invest 2001; 107(6):R37-R45.
60. Ross R, Reske-Kune AB. The role of NO in contact hypersensitivity. Int Immunopharmacol 2001; 1(8):1469-1478.
61. Waltz SE, Eaton L, Toney-Earley K et al. Ron-mediated cytoplasmic signaling is dispensable for viability but is required to limit inflammatory responses. J Clin Invest 2001; 108(4):567-576.
62. Lange-Asschenfeldt B, Weninger W, Velasco P et al. Increased and prolonged inflammation and angiogenesis in delayed-type hypersensitivity reactions elicited in the skin of thrombospondin-2-deficient mice. Blood 2002; 99(2):538-545.
63. Koyama Y, Kusubata M, Yoshiki A et al. Effect of tenascin-C deficiency of chemically induced dermatitis in the mouse. J Invest Dermatol 1998; 111(6):930-935.
64. Scholzen TE, Steinhoff M, Bonaccorsi P et al. Neutral endopeptidase terminates substance P-induced inflammation in allergic contact dermatitis. J Immunol 2001; 166(2):1285-1291.
65. Kadereit S, Xu H, Engeman TM et al. Negative regulation of CD8+ T cell function by the IFN-induced and double-stranded RNA-activated kinase PKR. J Immunol 2001; 165(12):6896-6901.

66. Duncas GS, Andrew DP, Takimoto H et al. Genetic evidence for functional redundancy of platelet/endothelial cell adhesion molecule-1 (PECAM-1): CD31-deficient mice reveal PECAM-1-dependent and PECAM-1-independent functions. J Immunol 1999; 162(5):3022-3030.
67. Gorbachev AV, Heeger PS, Fairchild RL. CD4+ and CD8+ T cell priming for contact hypersensitivity occurs independently of CD40-CD154 interactions. J Immunol 2001; 166(4):2323-2332.
68. Reeve VE, Bosnic M, Nishimura N. Interferon-γ is involved in photoimmunoprotection by UVA (320-400 nm) radiation in mice. J Invest Dermatol 1999; 112(6):945-950.
69. Zheng H, Fletcher D, Kozak W et al. Resistance to fever induction and impaired acute-phase response in interleukin-1β-deficient mice. Immunity 1995; 3(1):9-19.
70. Traidl C, Jugert F, Krieg T et al. Inhibition of allergic contact dermatitis to DNFB but not to oxazolone in interleukin-4-deficient mice. J Invest Dermatol 1999; 112(4):476-482.
71. Gorbachev AV, Dilulio NA, Fairchild RL et al. IL-12 augments $CD8^+$ T cell development for contact hypersensitivity responses and circumvents anti-CD154 antibody-mediated inhibition. J Immunol 2001; 167(1):156-162.
72. Byrum RS, Goulet JL, Snouwaert JN et al. Determination of the contribution of cysteinyl leukotrienes and leukotriene B4 in acute inflammatory responses using 5-lipoxygenase- and leukotriene A4 hydrolase-deficient mice. J Immunol 1999; 163(12):6810-6819.
73. Thomas JA, Allen JL, Tsen M et al. Impaired cytokine signaling in mice lacking the IL-1 receptor-associated kinase. J Immunol 1999; 163(2):978-984.
74. Miyauchi-Hashimoto H, Kuwamoto K, Urade Y et al. Carcinogen-induced inflammation and immunosuppression are enhanced in xeroderm pigmentosum group A model mice associated with hyperproduction of prostaglandin E_2. J Immunol 2001; 166(9):5782-5791.
75. Boonstra A, van Oudenaren A, Baert M et al. Differential ultraviolet-B-induced immunomodulation in XPA, XPC, and CSB DNA repair-deficient mice. J Invest Dermatol 2001; 117(1):141-146.
76. Garssen J, van Steeg H, de Gruijl F et al. Transcription-coupled and global genome repair differentially influence UV-B-induced acute skin effects and systemic immunosuppression. J Immunol 2000; 164(12):6199-6205.

CHAPTER 9

A Dual Role for Mast Cells in Contact Hypersensitivity Reactions:

More Team Players in Type 1 T Cell Mediated Contact Hypersensitivity Reactions

Tilo Biedermann and Martin Röcken

Summary

Contact hypersensitivity (CHS) reactions are prototypic delayed type hypersensitivity (DTH) reactions. They are mediated by interferon (IFN)-γ-producing $CD4^+$ and $CD8^+$, which are called type 1 T cells. Type 1 T cells can lead to the development of CHS reactions if directed against haptens or to autoimmune diseases when reacting with self antigens.[1-3] In humans, CHS reactions are T cell mediated skin diseases that display a heterogeneous inflammatory pattern depending on the nature of the contact allergen, the vehicle, and, most importantly, on the T cell and leukocyte subsets involved.[4-6] Consequently, the nature of murine and human CHS reactions as well as the underlying mechanisms are not identical, but share multiple similarities.[1,5] The most important common feature is the dependence on infiltrating T cells.[1,2,5,7] These effector T cells can directly mount destructive effector functions when activated in the skin.[5,8] To induce the full clinical picture of inflammation type 1 effector T cells have to attract and activate other leukocytes or resident cells within the peripheral tissue.[2] Several types of leukocytes are involved in extensive inflammation, but the induction of specific T cells that emigrate into the tissues is a necessary prerequisite for the accumulation of the different leukocytes. The exact orchestration leading to the recruitment and activation of the different cell populations in the skin is still unclear. Moreover, the initial events leading to extravasation of type 1 T cells into skin remained enigmatic. Some studies found an important role for a B cell product in CHS reactions, despite the obligatory role of T cells,[9] and morphologic studies suggested that activated mast cells (MC) play a role in human CHS reactions.[10,11] Recent findings now demonstrate that MC play a critical role both in the initiation and the amplification of CHS reactions. MC release the initial acute mediators opening the way for type 1 T cells to enter the skin and MC-derived factors later attract leukocytes like polymorphonuclear granulocytes (PMN) in response to T cell-derived signals.[2,12-15] Thus, MC provide the link for the B cell-dependent T cell recruitment and subsequently, MC respond to T cells during CHS.[2,9,15]

This chapter will present some of these most recent data on the role of MC in CHSR and discuss mechanisms, relevance, and some consequences of these findings.

Immune Mechanisms in Allergic Contact Dermatitis, edited by Andrea Cavani and Giampiero Girolomoni. ©2005 Eurekah.com.

The Dual Role of Mast Cells During Early and Late Phases of CHS

Mast cell (MC)-deficient $Kit^{W}/Kit^{W\text{-}v}$ mice can be primed to develop normal T cell-responses to haptens after immunization.[2,16] Nevertheless, T cell-dependent DTH like CHS reactions are significantly attenuated in $Kit^{W}/Kit^{W\text{-}v}$ mice,[2,13,14] showing that MC play an important role during the elicitation of CHS reactions. Earlier work has shown that CHS reactions are biphasic. Detailed analysis of CHSR in MC-deficient $Kit^{W}/Kit^{W\text{-}v}$ mice revealed that in these mice both the early phase after hapten challenge (up to 2 hours) as well as the late phase (around 24-48 hours) are strongly reduced.[2,13-15] Both, the early and the late reactions are T cell dependent.[12] Interestingly, in MC-deficient $Kit^{W}/Kit^{W\text{-}v}$ mice CHS reactions can partly be restored, if T cells are adoptively transferred directly into the skin, indicating that MC initiate T cell recruitment into the skin already during the early phase after hapten exposure.[13]

Mechanism of MC Activation in the Early Phase of CHS

MC need to be stimulated to secrete stored and newly synthesized mediators required to cause T cell extravasation into the skin. Specific hapten, present on the skin, can penetrate to the dermis like in contact urticaria. In contact urticaria, MC activation occurs after allergens crosslink IgE bound on the MC surface.[17] But, how do haptens activate MC to initiate CHS reactions? This putative role of MC in early CHS reactions has very recently been investigated.[15] Normal mice immunized with a hapten show specific recall CHS as early as 4 days after the immunization. In sharp contrast, B cell deficient mice display impaired CHS responses, characterized by a reduced early and late ear swelling and decreased production of interferon (IFN)-γ. Thus, B cells provide signals strictly required for early CHS development.[9,18] Since B cells need several days to even weeks to switch from IgM towards the production of other immunoglobulin subtypes, an antigen specific IgM was suspected to activate dermal MC after hapten exposure.[9,18] This can poorly explain MC activation, as pentameric IgM does not easily enter intercellular spaces. Early after activation, B cells may not only secrete complete antigen specific IgM, but also single light or heavy chains. These κ light single chains can specifically bind hapten antigen. It has now been shown that immunization with a hapten induces the production of hapten specific κ light chains,[15] which are capable of binding to a still unknown receptor on MC and—by encountering the specific hapten—mediate MC degranulation.[15] Activation of MC mediated by κ light chain-hapten complexes leads to MC-degranulation and possibly the secretion of a broad set of potent acute mediators, similar to allergen mediated IgE crosslinking on MC.[15,19] These MC derived factors should be capable of inducing the early changes required for the recruitment of type 1 T cells that induce CHS.[20]

Consequences of Early Antigen-Specific MC Activation

Recruitment of leukocytes into sites of inflammation requires two major events, attachment and migration.[21,22] Tumor necrosis factor (TNF) and interleukin (IL)-1 induce the expression of adhesion molecules on endothelial cells that are required for the attachment of lymphocytes and of PMN such as intercellular adhesion molecule 1 (ICAM-1) and vascular cell adhesion molecule 1 (VCAM-1).[23] Degranulation of MC in vivo can support different steps of leukocyte extravasation.[24] Since MC are the only cells carrying preformed TNF, it is intriguing to speculate that MC-degranulation can induce endothelial adhesion molecules,[19,25] and that MC derived TNF may be the most important mediator for leukocyte recruitment (Fig. 1, left). Consequently, while the reconstitution with wildtype MC completely restores CHS reactions in MC-deficient $Kit^{W}/Kit^{W\text{-}v}$ mice, reconstitution with TNF deficient MC does not.[2] In consequence, MC derived TNF may not only play a crucial role in the late phase but also in the early phase of CHS by initiating the recruitment of T lymphocytes and other leukocytes.[2,18] However, TNF induction alone is not sufficient for the recruitment of leukocytes as it requires specific chemoattractants in addition to TNF.

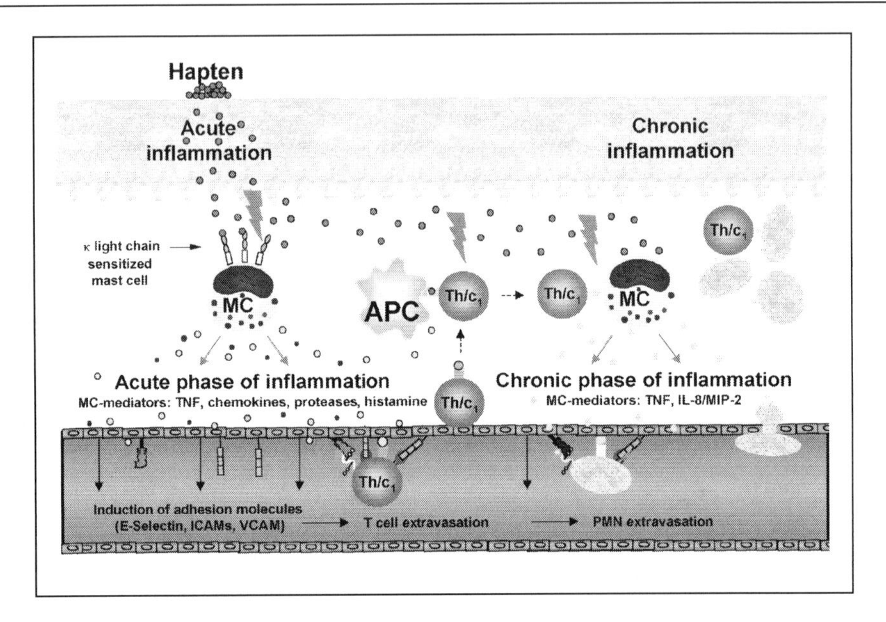

Figure 1. The dual role for mast cells in contact hypersensitivity reactions. Early activation of MC by antigen specific κ light chains encountering haptens induces the necessary molecules for T cell adhesion at the endothelial surface by MC derived TNF (left). MC activation also provides the chemotactic factors for T cell extravasation and trans-tissue migration (directly or indirectly). Following MC activation, type 1 T cells migrate into the dermis and encounter professional antigen presenting cells (APCs) carrying specific haptens (middle). Thereafter, activated type 1 T cells stimulate MC to specifically release TNF and MIP-2, which is a prerequisite for the infiltration of PMN, a hallmark of the late phase of CHS (right). Thus, MC tell T cells to emigrate from vessels during the initiation of CHS, but inversely respond to T cells during late phases.

Chemokines are a relatively new and still growing family of small molecular weight cytokines, which, among other functions, are responsible for the recruitment of leukocytes into sites of inflammation.[26,27] Specific chemokine chemokine-receptor interactions have been identified that are relevant exclusively for the homing of distinct leukocyte subsets. Inflammatory chemokines binding to CXCR3 and CCR5 as well as skin specific chemokines binding to CCR4 and CCR10 have been identified and are good candidates to mediate migration of type 1 T cells into skin during early CHS development.[2,27-32] Whether these chemokines are produced directly by activated MC, by keratinocytes, by activated dendritic cells, or other bystander cells still needs to be investigated.

Thus, it seems clear that early activation of MC by antigen specific κ light chains encountering haptens induces the necessary molecules for T cell adhesion at the endothelial surface by MC derived TNF (Fig. 1, left). Furthermore, MC early after activation may also provide the chemotactic factors for T cell extravasation and trans-tissue migration (directly or indirectly).

The Role of MC During Late Phases of CHS Reactions

Polymorphonuclear granulocytes (PMN) recruitment is a hallmark of late phase CHS. In MC-deficient $Kit^{W}/Kit^{W\text{-}v}$ mice, CHS reactions are virtually devoid of PMN infiltrates.[2,14] In infectious diseases, PMN are believed to be the first leukocytes that migrate into tissues in response to invading pathogens, while T cells seem to trigger a second wave of PMN influx in an antigen specific manner.[33] PMN are also present during inflammatory skin lesions such as CHS or psoriasis. It has long been recognized that T cells are preceding PMN during CHS, exactly the opposite to early inflammation during bacterial infection.[34,35] In such a way, PMN are recruited to inflammatory lesions in several T cell mediated autoimmune diseases including rheumatoid arthritis, Crohn's disease or inflammatory bowel disease. Interestingly, all these diseases are also associated with activated MC.[2,4,14,34,36-40] Since PMN recruitment is a hallmark of late phase CHS reactions, the mechanisms of PMN-recruitment during T cell-dependent CHS was analyzed in detail.[2] Specific chemokine chemokine-receptor interactions have been identified to be relevant for the homing of PMN. The growth-regulated oncogene (GRO) homologue KC and the functional analogue to human IL-8, macrophage inflammatory protein (MIP)-2, both bind to CXCR2 and are the most important chemokines for attracting PMN.[41] Thus, the chemokines GRO/KC and IL-8/MIP-2 are used as marker chemokines to investigate PMN involvement in inflammatory processes. Correlation studies revealed that selective induction of GRO/KC mRNA is associated with only sparse PMN accumulation.[2,42] Dense PMN infiltrates during type 1 T cell immune responses in CHS or rheumatoid arthritis correlate with the expression of both, GRO/KC and IL-8/MIP-2.[2,43] The crucial role of IL-8/MIP-2 for PMN extravasation was demonstrated by data showing that anti-MIP-2 antibodies block T cell-dependent PMN recruitment.[2] As a consequence, deviating type 1 T cell mediated responses into IL-4 dominated type 2 T cell reactions reduces the severity of inflammation, downregulates MIP-2 expression and almost completely prevents PMN infiltrates in affected skin sites.[44,45] This indicates that IL-8/MIP-2 expression, and PMN recruitment in CHS lesions directly depend on type 1 T cells, even though type 1 T cells produce only low amounts of IL-8/MIP-2.[46] Molecular and cellular analyses of CHS lesions unraveled that almost all MIP-2 required for PMN recruitment during CHS derived from dermal MC.[2] Reconstitution of MC-deficient $Kit^{W}/Kit^{W\text{-}v}$ mice with MC from wild-type and from tumor necrosis factor $(TNF)^{-/-}$ mice confirmed that MC are required for local IL-8/MIP-2 production and further showed that, in addition to IL-8/MIP-2, MC-derived TNF determines the extend of T cell mediated inflammation.[2] Thus, MC-derived TNF and MIP-2 are the mediators responsible for the changes characterizing the late phase of CHS reactions (Fig. 1).

Mechanism of MC Activation During Late Phases of CHS Reactions

These data raise the obvious question whether and how infiltrating type 1 T cells are able to activate dermal MC. Several studies have analyzed interactions of T cells and MC. Morphologic studies showed that activated MC can be found in close vicinity to T cells in many DTH reactions such as CHS or rheumatoid arthritis.[10,11,39] In fact, MC tend to form aggregates with activated T cells in an lymphocyte function-associated antigen (LFA)-1 dependent manner, which directly leads to histamine release.[47] Moreover, degranulation of MC through crosslinking FcεRI is augmented by coculture with activated, but not with resting T cells,[47] and activated T cells can induce MC to secrete the PMN attracting chemokine MIP-2/IL-8.[48] Under certain conditions, MC may also function as antigen presenting cells, at least in vitro, and inversely activate $CD4^{+}$ as well as $CD8^{+}$ T cells. T cell activation by MC depends on costimulatory signals and induces signaling in both, T cell and MC.[49-52] Such signaling must not necessarily cause MC degranulation, thus, type 1 T cells may also activate MC by soluble factors.[53] Importantly, type 1 T cell derived IFN-γ may induce other cytokines and chemokines in MC than Th2 derived factors.[54] Thus, distinct activation signals delivered by the various T cell cytokines could explain at least in part apparent differences seen, when either type 1 or type 2 T cells interact with MC.[45,55]

Some mechanisms of early MC activation—like hapten binding through antigen specific κ light chains to MC membranes—may also be involved in the late phase of CHS reactions.

Indeed, crosslinking of a still unknown receptor on MC through κ light chain-hapten complexes and activation of MC by type 1 T cells may synergize as shown for FcεRI-dependent MC degranulation.[47] This is supported by the finding that depletion of T cells completely abrogates late phase CHS pathology.[2] It is most likely that several players interact during the development of CHS, but that the cross talk between T cells and MC plays a crucial role. The data currently available support the concept that infiltrating type 1 T cells activate MC to specifically release TNF and MIP-2 through a still enigmatic mechanism and that this activation is a prerequisite for the infiltration of PMN, a hallmark of the late phase of CHS reactions. Thus, MC tell T cells to emigrate from vessels during the initiation of CHS, but inversely respond to T cells during late phases (Fig. 1).

The Dual Role for MC During Early and Late Phases of CHSR—A Unique or a General Phenomenon?

Arthus reactions and immune complex mediated inflammatory responses depend on MC activation, TNF secretion, and have a biphasic course. They develop an early edematous tissue reaction followed by a late phase with dominant leukocyte infiltration.[56-58] Similarily, allergen induced crosslinking of IgE-sensitized dermal MC results in early urtica, visible in prick test reactions, but may be followed by intense dermal leukocyte infiltrates consisting of T cells and PMN. Such IgE-initiated leukocyte recruitment depends largely on MC derived TNF.[19,59] A similar situation is given in asthma, where a MC and T cell dependent immune response starts as early phase events followed by PMN-rich late phase reactions.[55,60,61] Together, these findings suggest a general rule for acquired immune responses that involve MC. The initiating signals seem to be derived from antigen-specific B cells like IgE, immune-complexes, or κ light chains. They stimulate MC to cause the edema characterizing the early phase of inflammation. These immune responses then can develop late phases with sustained TNF expression. They depend in addition on T cell activation and lead to the recruitment of monocytes, PMN or eosinophilic granulocytes. These concepts may be extended to immune responses like Arthus reaction, immune-complex mediated inflammation, IgE mediated cutaneous late phase responses, and allergic asthma.[55-62]

Consequences of the Dual Role for MC in Acquired Immune Responses

It is intriguing to speculate that the findings on the dual role of MC in CHS reactions may also hold true in other harmful T cell mediated immune responses with MC involvement. Especially autoimmune diseases showing a pathology similar to CHS reactions including T cell and MC activation and PMN recruitment, may involve initial MC activation by κ light chains and late MC activation by T cells for PMN migration. These mechanistic steps in the development of CHS reactions may explain initiation and perpetuation of relapsing autoimmune diseases such as rheumatoid arthritis, Crohn's disease, inflammatory bowel disease (IBD) or psoriasis.[2,36-40] Thus, preventing the initiation of DTH reactions by inhibiting MC-activation through κ light chains or other B cell-derived mediators is an intriguing new tool for preventing CHS and possibly also other harmful DTH reactions.

References

1. Grabbe S, Schwarz T. Immunoregulatory mechanisms involved in elicitation of allergic contact hypersensitivity. Immunol Today 1998; 19:37-44.
2. Biedermann T, Kneilling M, Mailhammer R et al. Mast cells control neutrophil recruitment during T cell-mediated delayed-type hypersensitivity reactions through tumor necrosis factor and macrophage inflammatory protein 2. J Exp Med 2000; 192:1441-1452.
3. Rocken M, Biedermann T. Pathogenesis of autoimmune diseases. In: Hertl M, ed. Autoimmune diseases of the skin. Wien-New York: Springer, 2001:1-20.
4. Lerche A, Bisgaard H, Christensen JD et al. Human leukocyte mobilization and morphology in nickel contact allergy using a skin chamber technique. Acta Derm Venereol 1981; 61:517-523.
5. Girolomoni G, Sebastiani S, Albanesi C et al. T-cell subpopulations in the development of atopic and contact allergy. Curr Opin Immunol 2001; 13:733-737.
6. Kimber I, Dearman RJ. Allergic contact dermatitis: The cellular effectors. Contact Dermatitis 2002; 46:1-5.
7. Rocken M, Racke M, Shevach EM. IL-4-induced immune deviation as antigen-specific therapy for inflammatory autoimmune disease. Immunol Today 1996; 17:225-231.
8. Traidl C, Sebastiani S, Albanesi C et al. Disparate cytotoxic activity of nickel-specific CD8+ and CD4+ T cell subsets against keratinocytes. J Immunol 2000; 165:3058-3064.
9. Tsuji RF, Geba GP, Wang Y et al. Required early complement activation in contact sensitivity with generation of local C5-dependent chemotactic activity, and late T cell interferon gamma: A possible initiating role of B cells. J Exp Med 1997; 186:1015-1026.
10. Dvorak AM, Mihm Jr MC, Dvorak HF. Morphology of delayed-type hypersensitivity reactions in man. II. Ultrastructural alterations affecting the microvasculature and the tissue mast cells. Lab Invest 1976; 34:179-191.
11. Waldorf HA, Walsh LJ, Schechter NM et al. Early cellular events in evolving cutaneous delayed hypersensitivity in humans. Am J Pathol 1991; 138:477-486.
12. van Loveren H, Meade R, Askenase PW. An early component of delayed-type hypersensitivity mediated by T cells and mast cells. J Exp Med 1983; 157:1604-1617.
13. Askenase PW, Van Loveren H, Kraeuter Kops S et al. Defective elicitation of delayed-type hypersensitivity in W/Wv and SI/SId mast cell-deficient mice. J Immunol 1983; 131:2687-2694.
14. Webb EF, Tzimas MN, Newsholme SJ et al. Intralesional cytokines in chronic oxazolone-induced contact sensitivity suggest roles for tumor necrosis factor alpha and interleukin-4. J Invest Dermatol 1998; 111:86-92.
15. Redegeld FA, Van Der Heijden MW, Kool M et al. Immunoglobulin-free light chains elicit immediate hypersensitivity-like responses. Nat Med 2002; 8:696-703.
16. Galli SJ, Hammel I. Unequivocal delayed hypersensitivity in mast cell-deficient and beige mice. Science 1984; 226:710-713.
17. Wakelin SH. Contact urticaria. Clin Exp Dermatol 2001; 26:132-136.
18. Askenase PW, Tsuji RF. B-1 B cell IgM antibody initiates T cell elicitation of contact sensitivity. Curr Top Microbiol Immunol 2000; 252:171-177.
19. Wedemeyer J, Galli SJ. Mast cells and basophils in acquired immunity. Br Med Bull 2000; 56:936-955.
20. Rocken M, Hultner L. Heavy function for light chains. Nat Med 2002; 8:12-14.
21. Butcher EC, Picker LJ. Lymphocyte homing and homeostasis. Science 1996; 272:60-66.
22. Baggiolini M. Chemokines and leukocyte traffic. Nature 1998; 392:565-568.
23. McHale JF, Harari OA, Marshall D et al. TNF-alpha and IL-1 sequentially induce endothelial ICAM-1 and VCAM-1 expression in MRL/lpr lupus-prone mice. J Immunol 1999; 163:3993-4000.
24. Gaboury JP, Johnston B, Niu XF et al. Mechanisms underlying acute mast cell-induced leukocyte rolling and adhesion in vivo. J Immunol 1995; 154:804-813.
25. Walsh LJ, Trinchieri G, Waldorf HA et al. Human dermal mast cells contain and release tumor necrosis factor alpha, which induces endothelial leukocyte adhesion molecule 1. Proc Natl Acad Sci USA 1991; 88:4220-4224.
26. Baggiolini M, Loetscher P. Chemokines in inflammation and immunity. Immunol Today 2000; 21:418-420.

27. Kunkel EJ, Butcher EC. Chemokines and the tissue-specific migration of lymphocytes. Immunity 2002; 16:1-4.
28. Campbell JJ, Haraldsen G, Pan J et al. The chemokine receptor CCR4 in vascular recognition by cutaneous but not intestinal memory T cells. Nature 1999; 400:776-780.
29. Reiss Y, Proudfoot AE, Power CA et al. CC chemokine receptor (CCR)4 and the CCR10 ligand cutaneous T cell-attracting chemokine (CTACK) in lymphocyte trafficking to inflamed skin. J Exp Med 2001; 194:1541-1547.
30. Homey B, Alenius H, Muller A et al. CCL27-CCR10 interactions regulate T cell-mediated skin inflammation. Nat Med 2002; 8:157-165.
31. Kunkel EJ, Boisvert J, Murphy K et al. Expression of the chemokine receptors CCR4, CCR5, and CXCR3 by human tissue-infiltrating lymphocytes. Am J Pathol 2002; 160:347-355.
32. Sebastiani S, Albanesi C, Nasorri F et al. Nickel-specific CD4+ and CD8+ T cells display distinct migratory responses to chemokines produced during allergic contact dermatitis. J Invest Dermatol 2002; 118:1052-1058.
33. Appelberg R, Silva MT. T cell-dependent chronic neutrophilia during mycobacterial infections. Clin Exp Immunol 1989; 78:478-483.
34. Roupe G, Ridell B. The cellular infiltrate in contact hypersensitivity to picryl chloride in the mouse. Acta Derm Venereol 1979; 59:191-195.
35. Braun-Falco O, Schmoeckel C. The dermal inflammatory reaction in initial psoriatic lesions. Arch Dermatol Res 1977; 258:9-16.
36. Christophers E. The immunopathology of psoriasis. Int Arch Allergy Immunol 1996; 110:199-206.
37. Burmester GR, Daser A, Kamradt T et al. Immunology of reactive arthritides. Annu Rev Immunol 1995; 13:229-250.
38. Feldmann M, Brennan FM, Maini RN. Rheumatoid arthritis. Cell 1996; 85:307-310.
39. Malone DG, Irani AM, Schwartz LB et al. Mast cell numbers and histamine levels in synovial fluids from patients with diverse arthritides. Arthritis Rheum 1986; 29:956-963.
40. Strober W, Ehrhardt RO. Chronic intestinal inflammation: An unexpected outcome in cytokine or T cell receptor mutant mice. Cell 1993; 75:203-205.
41. Bozic CR, Kolakowski Jr LF, Gerard NP et al. Expression and biologic characterization of the murine chemokine KC. J Immunol 1995; 154:6048-6057.
42. Dilulio NA, Engeman T, Armstrong D et al. Groalpha-mediated recruitment of neutrophils is required for elicitation of contact hypersensitivity. Eur J Immunol 1999; 29:3485-3495.
43. Kasama T, Strieter RM, Lukacs NW et al. Interleukin-10 expression and chemokine regulation during the evolution of murine type II collagen-induced arthritis. J Clin Invest 1995; 95:2868-2876.
44. Biedermann T, Ogilvie A, Mai A et al. Reduced neutrophil chemotaxis following IL-4 therapy of established contact hypersensitivity (abstract). J Invest Dermatol 1997; 108:88.
45. Biedermann T, Mailhammer R, Mai A et al. Reversal of established delayed type hypersensitivity reactions following therapy with IL-4 or antigen-specific Th2 cells. Eur J Immunol 2001; 31:1582-1591.
46. Conlon K, Lloyd A, Chattopadhyay U et al. CD8+ and CD45RA+ human peripheral blood lymphocytes are potent sources of macrophage inflammatory protein 1 alpha, interleukin-8 and RANTES. Eur J Immunol 1995; 25:751-756.
47. Inamura N, Mekori YA, Bhattacharyya SP et al. Induction and enhancement of Fc(epsilon)RI-dependent mast cell degranulation following coculture with activated T cells: dependency on ICAM-1- and leukocyte function-associated antigen (LFA)-1-mediated heterotypic aggregation. J Immunol 1998; 160:4026-4033.
48. Krishnaswamy G, Lakshman T, Miller AR et al. Multifunctional cytokine expression by human mast cells: Regulation by T cell membrane contact and glucocorticoids. J Interferon Cytokine Res 1997; 17:167-176.
49. Fox CC, Jewell SD, Whitacre CC. Rat peritoneal mast cells present antigen to a PPD-specific T cell line. Cell Immunol 1994; 158:253-264.
50. Frandji P, Tkaczyk C, Oskeritzian C et al. Exogenous and endogenous antigens are differentially presented by mast cells to CD4+ T lymphocytes. Eur J Immunol 1996; 26:2517-2528.
51. Malaviya R, Twesten NJ, Ross EA et al. Mast cells process bacterial Ags through a phagocytic route for class I MHC presentation to T cells. J Immunol 1996; 156:1490-1496.

52. Frandji P, Mourad W, Tkaczyk C et al. IL-4 mRNA transcription is induced in mouse bone marrow-derived mast cells through an MHC class II-dependent signaling pathway. Eur J Immunol 1998; 28:844-854.
53. Leal-Berumen I, Conlon P, Marshall JS. IL-6 production by rat peritoneal mast cells is not necessarily preceded by histamine release and can be induced by bacterial lipopolysaccharide. J Immunol 1994; 152:5468-5476.
54. Stout RD, Bottomly K. Antigen-specific activation of effector macrophages by IFN-gamma producing (TH1) T cell clones. Failure of IL-4-producing (TH2) T cell clones to activate effector function in macrophages. J Immunol 1989; 142:760-765.
55. Lukacs NW, Hogaboam CM, Kunkel SL et al. Mast cells produce ENA-78, which can function as a potent neutrophil chemoattractant during allergic airway inflammation. J Leukoc Biol 1998; 63:746-751.
56. Zhang Y, Ramos BF, Jakschik BA. Neutrophil recruitment by tumor necrosis factor from mast cells in immune complex peritonitis. Science 1992; 258:1957-1959.
57. Ramos BF, Zhang Y, Jakschik BA. Neutrophil elicitation in the reverse passive Arthus reaction. Complement-dependent and -independent mast cell involvement. J Immunol 1994; 152:1380-1384.
58. Zhang Y, Ramos BF, Jakschik B et al. Interleukin 8 and mast cell-generated tumor necrosis factor-alpha in neutrophil recruitment. Inflammation 1995; 19:119-132.
59. Wershil BK, Wang ZS, Gordon JR et al. Recruitment of neutrophils during IgE-dependent cutaneous late phase reactions in the mouse is mast cell-dependent. Partial inhibition of the reaction with antiserum against tumor necrosis factor-alpha. J Clin Invest 1991; 87:446-453.
60. Galli SJ. New concepts about the mast cell. N Engl J Med 1993; 328:257-265.
61. Wills-Karp M. Immunologic basis of antigen-induced airway hyperresponsiveness. Annu Rev Immunol 1999; 17:255-281.
62. Gordon JR, Galli SJ. Release of both preformed and newly synthesized tumor necrosis factor alpha (TNF-alpha)/cachectin by mouse mast cells stimulated via the Fc epsilon RI. A mechanism for the sustained action of mast cell-derived TNF-alpha during IgE-dependent biological responses. J Exp Med 1991; 174:103-107.

Chapter 10

Multiple Roles of Keratinocytes in Allergic Contact Dermatitis

Cristina Albanesi, Andrea Cavani, Claudia Scarponi and Giampiero Girolomoni

Summary

Cellular and molecular mechanisms underling allergic contact dermatitis (ACD) have been, and still are, under intense investigation, not only for the relevance of ACD in clinical medicine, but also because ACD is a paradigm of T cell-mediated immune reactions of the skin. Epidermal cells are far more than mere spectators of these immune responses, and undergo a functional activation after contact with haptens and subsequently with lymphokines, causing ACD reactions. Keratinocyte activation includes the expression of a plethora of cytokines, chemokines and accessory molecules, which can profoundly influence immunocompetent cells, thereby contributing to the dramatic alterations that occur during ACD. Lastly, activated keratinocytes also produce and become susceptible to immunosuppressive cytokines which can dampen ACD reactions.

Introduction

The primary function of keratinocytes is to provide the structural and physical integrity of the skin, thereby protecting the body against external injury. Although in the past keratinocytes were considered simply as passive targets of immunological attack from infiltrating T lymphocytes, a number of studies have definitively demonstrated that keratinocytes importantly participate in cutaneous immune responses by transmitting both positive and negative signals to cells of innate and adaptive immunity.[1,2] As an inevitable counterpart of its location and duty to prepare efficient immune responses, the skin is a frequent site of hypersensitivity reactions against apparently harmless antigens, such as the haptens causing allergic contact dermatitis (ACD). Valuable animal models of ACD are available, and the disease can be easily induced in humans, allowing a detailed analysis of the cell types and molecules involved in its initiation, amplification and regulation. Over the past decade, numerous observations have detailed the nature of keratinocyte involvement in both the initiation and amplification phases of ACD reactions.[3,4] In particular, reactive haptens or T-cell-derived lymphokines have been identified as triggers of keratinocyte expression of soluble and membrane inflammatory mediators, which are directly and indirectly responsible for the recruitment, and local activation of T cells and other leukocytes.[5-7] On the other hand, keratinocytes possess the capacity to produce immuno-suppressive factors and cytokines involved in the inhibition of

Immune Mechanisms in Allergic Contact Dermatitis, edited by Andrea Cavani and Giampiero Girolomoni. ©2005 Eurekah.com.

the pro-inflammatory responses mediated by keratinocytes themselves or other genuine immune skin cells during ACD.[8-9]

Hapten-Activated Keratinocytes As Initiators of ACD Reactions

ACD reactions typically develop in two sequential phases: sensitization and elicitation. In the sensitization phase, haptens are collected by skin dendritic cells, carried to regional lymph nodes and presented to naive T cells. Re-exposure to causal hapten initiates the elicitation phase during which memory hapten-specific T cells recirculating in the skin are rapidly activated and express their effector functions, and ultimately cause tissue injury. Resident keratinocytes actively regulate both the sensitization and elicitation phases of ACD responses. In fact, several studies conducted in vivo and in vitro on cultures of human keratinocyte and keratinocyte-like cell lines demonstrated that haptens such as urushiol and nickel induce a direct inflammatory activation of keratinocytes resulting in the expression of cytokines, chemokines and adhesion molecules.[3,10-11] Among keratinocyte-derived cytokines, tumor necrosis factor (TNF)-α, interleukin (IL)-1 and granulocyte macrophage colony stimulating factor (GM-CSF) have pleiotropic effects and can be considered as primary promoting factors of the cutaneous allergic responses to haptens given their own capacity to activate various cell populations of the skin. For instance, under the influence of IL-1 and GM-CSF, Langerhans cells (LCs) acquire higher immunostimulatory capacities whereas IL-1 and TNF-α are important signals for their emigration from the epidermis to the draining lymph nodes.[12-13] TNF-α and IL-1 are also potent activating stimuli for dermal endothelial cells, which in turn can conceivably contribute to leukocyte recruitment and arrival in the skin by expressing both adhesion molecules (e.g., selectins, intercellular adhesion molecule (ICAM)-1 and vascular cell adhesion molecule (VCAM)-1) and chemokines (e.g., IL-8/CXCL8 and macrophage inflammatory protein (MIP)-3α/CCL20).[14,15] Moreover, both these cytokines strongly stimulate endothelial cells to generate platelet activating factor which is instrumental in the process of early neutrophil migration across the endothelium.[16] Finally, TNF-α released by hapten-activated keratinocytes can act in an autocrine manner on keratinocytes themselves and is responsible for ICAM-1 and IL-8/CXCL8 expression observed in the early phase of ACD prior to dermal leukocyte infiltration.[17] TNF-α relevance in ACD has been extensively studied also in genetically manipulated mice: induction of contact hypersensitivity (CHS) response to haptens is depressed in mice lacking TNF-α or the TNF-R2, whereas, TNF-R1 knock-out mice have enhanced responses.[18]

In the early phase of ACD, IL-8/CXCL8 and the related chemokine growth-regulated oncogene (GRO)-α/CXCL1 are expressed by keratinocytes at moderate levels and are possibly responsible for the recruitment of the few neutrophils detected in early ACD lesions. In addition, recent data obtained in the murine system suggested that Gro-α/CXCL1 expression and subsequent neutrophil infiltration are prerequisites for T cell recruitment and elicitation of CHS reaction.[19] Monocyte chemoattractant protein (MCP)-1/CCL2 is also expressed by basal layer keratinocytes directly following their activation with contact allergen, as assessed by in situ hybridization performed on biopsies taken from ACD patients at different time points after hapten application.[20] MCP-1/CCL2 expression appeared at 6 h after hapten challenge and anticipated the infiltration of $CD3^+$ T lymphocytes and $CD68^+$ macrophages, which occurred at the time interval of 12 h and reached maximum intensities at 48 to 72 h. Similarly, cutaneous T cell-attracting chemokine (CTACK)/CCL27 is strongly expressed by epidermal keratinocytes of basal and suprabasal layers 6 h after nickel exposure, with this expression co-localizing with CCR10 perivascular reactivity.[21] Taken together these data indicate that, in the initial phase of ACD reaction, basal keratinocytes are important producers of inflammatory mediators and activators of other resident skin cells. At this stage, T cells and other leukocyte subpopulations have not been recruited to the skin, supporting the concept that keratinocytes are directly activated by contact allergens rather than stimulated by cytokine-releasing infiltrating cells. Indeed, Grabbe and colleagues demonstrated that the non specific effects of epicutaneously applied haptens contribute to the elicitation of CHS and that the irritative rather than antigen-specific properties of the hapten are responsible for the strict concentration-dependence of the effector phase of CHS.[22]

Critical Role of Keratinocytes in the Amplification Phase of ACD

In the elicitation phase of ACD, the initiation of allergic reactions is followed by the amplification of cutaneous inflammation, with increased sequestration of circulating leukocytes and accumulation of these cells in the skin at both dermal and epidermal levels.[23,24] Early (12-24 h) ACD lesions are characterized by the presence of a massive infiltrate of hapten-specific $CD4^+$ and $CD8^+$ T lymphocytes, which are activated and expanded following their interaction with hapten-activated skin dendritic cells. Among hapten-specific T lymphocytes, IFN-γ-secreting type 1 T clones predominate although a substantial proportion of IL-4-releasing type 2 lymphocytes are also present.[23,24] In addition, both Th1 and Th2 subsets release TNF-α and IL-17.[6,25] Many in vitro studies have indicated that T cell-derived cytokines are the most potent stimuli for keratinocyte activation. In particular, IFN-γ is the most potent activator of the pro-inflammatory functions of keratinocytes, which constitutively express the IFN-γ receptor complex.[17,25] ICAM-1 antigen is highly induced by IFN-γ in keratinocytes and has the important role of mediating the compartmentalization of LFA-1-bearing T lymphocytes in the epidermis during ACD.[26,27] In addition, ICAM-1 can serve as a relevant accessory signal for activation of both $CD4^+$ and $CD8^+$ T lymphocytes. In ACD skin, keratinocyte ICAM-1 is markedly increased and correlates with a parallel large amount of soluble ICAM-1, a form that derives from proteolytic cleavage of the membrane-bound form and that promote de-adhesion between leukocytes and ICAM-1-positive cells.[28] Moreover, the cytokines TNF-α, IL-17 and IL-4 reinforce the induction of ICAM-1 promoted by IFN-γ.[6,17,25] It has been amply documented that keratinocytes can express MHC class II molecules during ACD reactions, and this phenomenon has been attributed to IFN-γ released by infiltrating Th0 or Th1 cells.[29,30] MHC class II expression can be reproduced in vitro on cultured keratinocytes upon treatment with IFN-γ, which also induces the expression of accessory molecules responsible for the maturation of class II dimers.[5] Following IFN-γ treatment, keratinocytes express in fact HLA-DM and invariant chain genes under the control of CIITA and RFX5 transcription factors, and MHC class II molecules show a remarkable resistance to detergents, a feature that class II molecules acquire when their groove is properly loaded with peptide. These results suggest that human keratinocytes activated with IFN-γ possess the biochemical requirements for the generation of functional class II-peptide complexes. As a consequences, keratinocytes exposed to IFN-γ become susceptible to Th1-mediated cytotoxicity.[31] Keratinocytes constitutively express MHC class I and Fas molecules, which are only up-regulated by IFN-γ.[31,32] Therefore, $CD8^+$ T cells can exert their cytolytic activity on both resting and IFN-γ-activated keratinocytes. Antigen-loaded keratinocytes can be thus target of T cell-mediated cytotoxicity, with $CD8^+$ being responsible for the initiation of epidermal damage during ACD, and $CD4^+$ T cells cooperating with $CD8^+$ T cells at later times in causing tissue damage.

Other than T cell costimulatory molecules, keratinocytes produce a number of cytokines important in the expression of ACD. During a patch test reaction, keratinocytes release increased amounts of IL-1.[33] Both IL-1α and IL-1β are produced and stored within keratinocytes in large amounts and are easily released following cell damage. IL-1 can act on keratinocyte themselves and stimulate the synthesis of chemokines and other cytokines.[34] IL-1 also strengthens the antigen presenting capacity of dendritic cells to T lymphocytes, and induces adhesion molecules on endothelial cells.[35] Keratinocytes in vitro and intact skin in vivo, contain an intracellular variant of IL-1 receptor antagonist (IL-1ra).[36] IL-1ra is structurally related to IL-1α and IL-1β and binds to IL-1 receptors on various target cells including keratinocytes, without inducing any biological response.[37] Not only IL-1 bioactivity is regulated by specific receptor antagonism, but it is also modulated by expression of two species of surface receptors both expressed by keratinocytes, IL-1R type I and type II, with the type II not being able to mediate a signal transduction, but rather acting as an IL-1 scavenger.[38] Experimental evidence has clearly shown that IL-1ra is an effective inhibitor of both the sensitization and elicitation phases of CHS in the mouse.[8] An imbalance in IL-1α and IL-1ra production has important consequences on the inflammatory potential of IL-1, and the ratio of intracellular IL-1ra to IL-1α concentration is known to diminish in keratinocytes after IFN-γ, IL-4 or IL-17 treatment.[25] A dysregulation of the IL-1 system has been extensively documented in psoriasis, and is thought to be important in many other inflammatory skin diseases, including IgE-mediated allergic disorders and CHS responses.[39] Exposure to IL-1 stimulates keratinocyte release of a number of different cytokines, including IL-6 and GM-CSF.[1] CHS reactions are associated with a rapid and substantial increase in the plasma concentration of IL-6, and histochemical studies on human allergic patch test reactions confirmed enhanced keratinocyte expression of IL-6.[40] On the other hand, subcutaneous administration of IL-6 at the site of the challenge significantly suppressed both the induction and the elicitation phases of delayed-type reaction to protein antigens, suggesting that this cytokine may rather exert a regulatory function in this type of reaction.[41] GM-CSF is an another important inflammatory mediator which is produced by keratinocytes in pathologic conditions, known to be essential for dendritic cell development and deeply involved in the regulation of dendritic cell functions.[42] GM-CSF is a strong mitogenic stimulus for human keratinocytes, thus representing a favored autocrine factor responsible for the epidermal hyperplasia observed in ACD skin.[43] Both IL-6 and GM-CSF release by keratinocytes is finely regulated by a number of cytokines, including IFN-γ, TNF-α, IL-17 and IL-4.[25] During ACD, cytokine-activated keratinocytes become also an important source of chemotactic factors that direct the recruitment of specific leukocyte subpopulations. Numerous in situ studies on ACD skin have documented the increased keratinocyte expression of mRNA and/or protein for several chemokines, which are released in a sequential and coordinated manner. During ACD reaction, concomitantly to the early immigration of leukocytes and their release of inflammatory cytokines, very high levels of regulated upon activation, normal T cell expressed and secreted (RANTES)/CCL5, IFN-induced protein (IP)-10/CXCL10, monokine induced by IFN-gamma (Mig)/CXCL9, IFN-inducible T cell alpha chemoattractant (I-TAC)/CXCL11, MCP-1/CCL2 and MIP-3α/CCL20 are expressed by activated keratinocytes.[2,7,20,44] The expression of RANTES/CCL5, IP-10/CXCL10 and Mig/CXCL9 begins at 12 h after hapten application and reaches the maximum at 72 h, paralleling the strong infiltration of lymphocytes. IP-10/CXCL10, Mig/CXCL9 and I-TAC/CXCL11 are the chemokines which are more abundantly produced by activated keratinocytes, with the relevance of CXCR3 agonists residing in the fact that >70% of infiltrating cells express CXCR3 in ACD skin.[7,20] IFN-γ and TNF-α strongly induce IP-10/CXCL10, Mig/CXCL9 and I-TAC/CXCL11 expression in keratinocytes and, notably, IL-4, a type 2 cytokine, enhances IFN-γ- and TNF-α-induced release of CXCR3 agonists, indicating that the Th2 to Th1 switch observed in some chronic inflammatory skin diseases (e.g., atopic dermatitis) could partly depend on the predilection of keratinocytes to release Th1-active chemokines.[7] In vitro studies performed on normal human keratinocytes activated with nickel-specific T cell-derived supernatants, demonstrate that keratinocytes are more sensitive to Th1- than to Th2-derived lymphokines in terms of variety and amounts of chemokine released and promote the preferential migration of Th1 lymphocytes.[45] Among keratinocyte-derived chemokines, MCP-1/CCL2 also seems to play a relevant role in ACD. Transgenic mice overexpressing MCP-1/CCL2 in basal keratinocytes show enhanced CHS responses together with an increased number of infiltrating dendritic cells.[46] Moreover, following activation with IFN-γ and TNF-α, keratinocytes rapidly up-regulates IL-8/CXCL8 and CTACK/CCL27, and synthesize Gro-α/CXCL1, MIP-3α/CCL20, MDC/CCL22 and I-309/CCL1, with the latter two chemokines being produced at lower levels and with a delayed kinetics.[21,45,47] Differently from other chemokines which are widely and redundantly produced by a variety of cell types, CTACK/CCL27 is produced only by keratinocytes, and in ACD lesions, likewise psoriasis and atopic dermatitis, is highly represented.[21] During the elicitation of ACD, CTACK/CCL27 immunoreactivity is detectable in the epidermis and also free in the papillary dermis and on dermal microvessels, most likely as the consequences of secretion from keratinocytes followed by absorption on extracellular matrix proteins and endothelium, respectively. CTACK/CCL27 expression in ACD skin correlates with the increased number of $CCR10^{+}$ T cells first in the perivascular and subepidermal areas and then in the epidermis. CTACK/CCL27 expression in keratinocytes is primarily induced by TNF-α and IL-1β, and not by IFN-γ or IL-4.[21]

It is clear that keratinocytes act as potent "signal transducers", capable of converting proinflammatory lymphokine stimuli into production and release of a vast array of multifunctional cytokines, chemokines and membrane molecules involved in the amplification of ACD reactions. The blockade of expression of keratinocyte-derived inflammatory mediators would be advantageous for the resolution of ACD immune reaction and this aspect is discussed in the upcoming sections.

Contribution of Keratinocytes to the Suppression of ACD Reactions

Intervention of regulatory mechanisms to limit excessive tissue damage and promote the termination of ACD is essential for maintaining skin integrity. Cytokines are profoundly implicated in the modulation of immune responses and IL-10 has been shown to suppress murine CHS potently. Nickel-specific $CD4^{+}$ T cells that produce high levels of IL-10 have been isolated from allergic individuals and shown to block in an IL-10-dependent manner the maturation of dendritic cells and their ability to activate Tc1 and Th1 effector lymphocytes.[24,48] IL-10 production has been detected also in keratinocyte cultures after ultraviolet B radiation and may play an important role in ultraviolet-induced immunosuppression.[49] However, this immunosuppressive cytokine has been found only in murine keratinocyte strains and not in human keratinocytes nor epidermal cell lines, even after treatment with ultraviolet or a variety of other stimuli.[50] Moreover, it has been demonstrated that circulating IL-10 released by gene-transferred keratinocytes inhibits the effector phase of CHS at a distant area of the skin.[51] IL-10 inhibitory effects on CHS response are mainly explicated on antigen presenting cells, including epidermal LCs. In particular, IL-10 is shown to convert LCs from potent inducers of primary immune responses to hapten-specific tolerizing cells, both in vitro and in vivo.[52] Other than on dendritic cells, the IL-10 receptor (IL-10R) is expressed and functional also on keratinocytes, where it mediates a reduction of growth rate and inhibition of IFN-γ-induced HLA-DR expression.[53] IL-10 in situ binding has been shown in chronic psoriatic skin and IL-10R gene expression can be down-modulated by IL-8/CXCL8 and increased by corticosteroids or vitamin D3.[53] In contrast to these data, Seifert et al failed to detect any considerable effect of IL-10 on both unstimulated and IFN-γ- or LPS-activated primary human keratinocytes.[54] Recent data from our Laboratory demonstrated that keratinocytes express constitutively only the IL-10R β-chain, the subunit of IL-10R complex involved mainly in the IL-10 signal transduction pathway. Upon a prolonged (> 36 h) treatment with IFN-γ, keratinocytes up-regulate the IL-10R α-chain, the subunit responsible for the IL-10 binding to the IL-10R complex, and become functionally susceptible to the IL-10 anti-inflammatory effects (unpublished observations). Resting and activated keratinocytes can also produced transforming growth factor (TGF)-β, a cytokine with a known down-regulatory activity on immune processes.[9] In the mouse, systemic administration of recombinant TGF-β completely prevented CHS when the cytokine was administered during elicitation, whereas systemic administration did not affect the sensitization phase.[55] TGF-β anti-inflammatory property can be partly due to its ability to induce IL-1ra in different cell types.[56]

As mentioned in the previous section, IFN-γ-activated keratinocytes express MHC class II as well as the genes necessary for the production of peptide-loaded MHC class II molecules. However, keratinocytes fail to express adequate levels of costimulatory molecules such as CD80 and CD86, essential signals for a proper and strong activation of naive and resting T cells.[57] Lacking of CD80/CD86 expression by keratinocytes can explain previous studies showing that murine Ia$^+$ keratinocytes do not activate antigen-specific T cell proliferation and lymphokine production but, instead, they induce clonal anergy.[58] Moreover, using mouse model of in vivo antigen presentation for CHS Ia$^+$ keratinocytes are also tolerogenic.[59] Thus, it is logical to speculate that the induced expression of MHC class II molecules on keratinocytes may serve a critical role in down-regulating ACD reactions in the skin. That is, functional antigen processing cells, such as class II$^+$ LCs, must compete with non functional class II$^+$ keratinocytes to present antigen to hapten-specific T cells. It is noteworthy that keratinocytes greatly outnumber LCs in the epidermis and that the number of LCs in the epidermis decreases after allergen application. Therefore, keratinocytes may be deeply involved in the termination of the T cell-mediated inflammatory responses during ACD by inducing T cell clonal anergy.

Concluding Remarks

There is increasing evidence that the skin provides a complex microenviroment for the initiation and shaping of T cell-mediated immune responses. In particular, keratinocytes can elaborate exogenous and endogenous stimuli into a network of modulators of inflammation and immunity, which are relevant in the induction and expression of ACD reactions. Among these stimuli, IFN-γ is the most potent activator of the pro-inflammatory functions of keratinocytes, and its blockade could be an important strategy for the control of ACD. Recent findings from our Laboratory have demonstrated that the expression of membrane ICAM-1 and HLA-DR, and the release of IP-10/CXCL10, Mig/CXCL9 and MCP-1/CCL2 induced by IFN-γ in human keratinocytes can be abrogated by overexpressing the negative regulators of IFN-γ signaling SOCS1 and SOCS3.[60] SOCS (from suppressor of cytokine signaling) are a family of intracellular proteins which regulate the magnitude and duration of responses triggered by various cytokine by inhibiting their signal transduction pathway in a classic negative feedback loop.[61] SOCS molecules are strongly expressed by keratinocytes in Th1-mediated skin disorders such as ACD and psoriasis, and to a lesser extent in the Th2-dominated disease atopic dermatitis.[60] The inhibition of the IFN-γ-induced IP-10/CXCL10, Mig/CXCL9 and MCP-1/CCL2 observed in keratinocytes permanently transduced with SOCS1 and SOCS3 is dependent on the reduction of the IFN-γRα phosphorylation and, consequently, of signal transducer and activator of transcription (STAT)1 and STAT3 protein activities.[60] These findings identify SOCS1 and SOCS3 molecules as new potential molecular targets for those IFN-γ-dependent skin diseases where pro-inflammatory mediators are aberrantly expressed by keratinocytes.

Several clues have been presented in this review, demonstrating that in the skin complex mechanisms are on set both for the promotion and down-modulation of inflammatory events, whose complexity is progressively receiving substantial clarification. Their understanding will permit the designation of novel preventive and therapeutic strategies.

References

1. Pastore S, Cavani A, Girolomoni G. Epidermal and neuronal peptide modulation of contact hypersensitivity reactions. Immunopharmacology 1996; 31:117-130.
2. Sebastiani S, Albanesi C, De Pità O et al. The role of chemokines in allergic contact dermatitis. Arch. Dermatol Res 2002; 293:552-559.
3. Barker JN, Mitra RS, Griffiths CM et al. Keratinocytes as initiator of inflammation. Lancet 1991; 337:211-214.
4. Barker JN. Role of keratinocytes in allergic contact dermatitis. Contact Dermatitis 1992; 26:145-148.
5. Albanesi C, Cavani A, Girolomoni G. Interferon-γ-stimulated human keratinocytes express the genes necessary for the production of peptide-loaded MHC class II molecules. J Invest Dermatol 1998; 110:138-142.
6. Albanesi C, Cavani A, Girolomoni G. IL-17 is produced by nickel-specific T lymphocytes and regulates ICAM-1 expression and chemokine production in human keratinocytes: synergistic or antagonist effects with IFN-γ and TNF-α. J Immunol 1999; 162:494-502.
7. Albanesi C, Scarponi C, Sebastiani S, Cavani A, Federici M, De Pità O, Puddu P, Girolomoni G. IL-4 enhances keratinocyte expression of CXCR3 agonistic chemokines. J Immunol 2000; 165:1395-1402.
8. Kondo S, Pastore S, Fujisawa H et al. Interleukin-1 receptor antagonist suppresses contact hypersensitivity. J Invest Dermatol 1995; 105:334-338.
9. Stadnyk AW. Cytokine production by epithelial cells. FASEB J 1994; 8:1041-1047.
10. Little MC, Metcalfe RA, Haycock JW et al. The participation of proliferative keratinocytes in the preimmune response to sensitizing agents. Br J dermatol 1998; 138:45-56.
11. Lisby S, Muller K M, Jongeneel C V et al. Nickel and skin irritants up-regulate tumor necrosis factor-alpha mRNA in keratinocytes by different but potentially synergistic mechanisms. Int Immunol 1995; 7:343-352.
12. Jonuleit H, Knop J, Enk AH. Cytokines and their effects on maturation, differentiation and migration of dendritic cells. Arch Dermatol Res 1996; 289:1-8.
13. Cumberbatch M, Kimber J. Dermal tumor necrosis factor-alpha induces dendritic cell migration to draining lymph nodes and possibly provides one stimulus for Langerhans cell migration. Immunology 1992; 75:257-263.
14. Nickoloff BJ. Role of epidermal keratinocytes as key initiators of contact dermatitis due to allergic sensitization and irritation. J Am Acad Dermatol 1992; 3:65-69.
15. Fitzhugh DJ, Shubhada N, Caughman SW et al. C-C chemokine receptor 6 is essential for arrest of a subset of memory T cells on activated dermal microvascular endothelial cells under physiologic flow conditions in vitro. J Immunol 2000; 165:6677-6681.
16. Kuijpers TW, Hakkert BC, Hart ML et al. Neutrophil migration across monolayers of cytokine-prestimulated endothelial cells: a role for platelet-activating factor and IL-8. J Cell boil 1992; 117:565-573.
17. Barker JN, Sarma V, Mitra RS et al. Marked synergism between TNF-α and IFN-γ in regulation of keratinocyte-derived adhesion molecules and chemotactic factors. J Clin Invest 1990; 85:605-608.
18. Wang B, Feliciani C, Freed I et al. Insights into molecular mechanisms of contact hypersensitivity gained from gene knockout studies. J Leukoc Biol 2001; 70:185-191.
19. Dilulio NA, Engeman T, Armstrong D et al. Groalpha-mediated recruitment of neutrophils is required for elicitation of contact hypersensitivity. Eur J Immunol 1999; 29:3485-3495.
20. Goebeler M, Trautmann A, Voss A et al. Differential and sequential expression of multiple chemokines during elicitation of allergic contact hypersensitivity. Am J Pathol 2001; 158:431-440.
21. Homey B, Alenius H, Muller A et al. CCL27-CCR10 interactions regulate T cell-mediated skin inflammation. Nature Med 2002; 8:157-165.
22. Grabbe S, Steinert M, Mahnke K et al. Dissection of antigenic and irritative effects of epicutaneously applied haptens in mice. Evidence that not the antigenic component but non-specific proinflammatory effects of haptens determine the concentration-dependent elitation of allergic contact dermatitis. J Clin Invest 1996; 98:1158-1164.
23. Cavani A, Albanesi C, Traidl C et al. Effector and regulatory T cells in allergic contact dermatitis. Trends Immunol 2001; 22:118-120.

24. Girolomoni G, Sebastiani S, Albanesi C et al. T-cell subpopulations in the development of atopic and contact allergy. Curr Opin Immunol 2001; 13:733-737.
25. Albanesi C, Scarponi C, Federici F et al. Interleukin 17 is produced by both Th1 and Th2 lymphocytes, and modulates interferon-γ- and interleukin-4-induced activation of human keratinocytes. J Invest Dermatol 2000; 115:81-87.
26. Dustin ML, Singer KH, Tuck DT et al. Adhesion of T lymphoblasts to epidermal keratinocytes is regulated by interferon-γ and is mediated by intercellular adhesion molecule-1 (ICAM-1). J Exp Med 1988; 167:1323-1330.
27. Garioch JJ, Mackie RM, Campbell I et al. Keratinocyte expression of intercellular adhesion molecule 1 (ICAM-1) correlates with infiltration of lymphocyte function associated antigen 1 (LFA-1) positive cells in evolving allergic contact dermatitis reactions. Histopathology 1991; 18:351-356.
28. Ballmer-Weber BK, Braathen LR, Brand CU. sICAM-1, sE-selectin and sVCAM-1 are constitutively present in human skin lymph and increased in allergic contact dermatitis. Arch Dermatol Res 1997; 289:251-256.
29. Aubock J, Romani N, Grubauer G et al. HLA-DR expression on keratinocytes is a common feature of diseased skin. Br J Dermatol 1986; 115:465-472.
30. Wikner NE, Huff CJ, Norris DA et al. Study of HLA-DR synthesis in cultured human keratinocytes. J Invest Dermatol 1986; 87:559-564.
31. Traidl C, Sebastiani S, Albanesi C et al. Disparate cytotoxic activity of nickel-specific $CD8^+$ and $CD4^+$ T cell subsets against keratinocytes. J Immunol 2000; 165:3058-3064.
32. Sayama K, Yonehara S, Watanabe Y et al. Expression of Fas-antigen on keratinocytes in vivo and induction of apoptosis in cultured human keratinocytes. J Invest Dermatol 1994; 103:330-338.
33. Larsen CG, Ternowitz T, Larsen FG et al. Epidermis and lymphocyte interactions during an allergic patch test reaction: increased activity of ETAF/IL-1, epidermal derived lymphocyte chemotactic factor and mixed skin lymphocyte reactivity in persons with type IV allergy. I Invest Dermatol 1988; 90:230-233.
34. Kupper TS, Lee F, Birchall N et al. Interleukin 1 binds to specific receptors on human keratinocytes and induces granulocyte macrophage colony-stimulating factor mRNA and protein: a potential autocrine role of interleukin 1 in epidermis. J Clin Invest 1988; 82:1787-1792.
35. Mizell SB. Interleukin 1 and T-cell activation. Immunol Today 1987; 8:330-332.
36. Bigler CF, Norris DA, Weston W L et al. Interleukin-1 receptor antagonist production by human keratinocytes. J Invest Dermatol 1992; 98:38.44.
37. Hannum CH, Wilcox CJ, Arend WP et al. Interleukin-1 receptor antagonist activity of human interleukin-1 inibitor. Nature 1990; 343:336-340.
38. Colotta F, Re F, Muzio M et al. Interleukin-1 type II receptor: A decoy target for IL-1 that is regulated by IL-4. Science 1993; 261:472-475.
39. Hammerberg C, Arend WP, Fisher GJ et al. Interleukin-1 receptor antagonist in normal and psoriatic epidermis. J Clin Invest 1992; 98:336-344.
40. Oxholm A, Oxholm P, Avnstorp C et al. Keratinocyte-expression of interleukin 6 but not of tumor necrosis factor-alpha is increased in the allergic and the irritant patch test reaction. Acta Derm Venereol (Stockh) 1991; 71:93-98.
41. Mihara M, Ikuta M, Koishihara Y et al. Interleukin 6 inhibits delayed-type hypersensitivity and the development of adjuvant arthritis. Eur J Immunol 1991; 21:2327-2331.
42. Pastore S, Fanales-Belasio E, Albanesi C et al. Granulocyte macrophage colony-stimulating factor is overproduced by keratinocytes in atopic dermatitis. J Clin Invest 1997; 99:3009-3017.
43. Braunstein S, Kaplan G, Gottlieb AB et al. GM-CSF activates regenerative epidemal growth and stimulates keratinocyte proliferation in human skin in vivo. J Invest Dermatol 1994; 103: 601-604.
44. Flier JD, Boorsma DM, Bruynzeel DP et al. The CXCR3 activating chemokine IP-10, Mig and IP-9 are expressed in allergic but not in irritant patche test reactions. J Invest Dermatol 1999; 113:574-561.
45. Albanesi C, Scarponi C, Sebastiani S et al. A cytokine-to-chemokine axis between T lymphocytes and keratinocytes can favor Th1 cell accumulation in chronic inflammatory skin diseases. J Leukoc Biol 2001; 70:617-623.

46. Nakamura K, Williams IR, Kupper TS. Keratinocyte-derived monocyte chemoattractant protein 1 (MCP-1): analysis in a transgenic model demonstrates that MCP-1 can recruit dendritic and Langerhans cells to skin. J Invest Dermatol 1995; 105:635-643.
47. Homey B, Dieu-Nosejan MC, Wiesenborn A et al. Up-regulation of macrophage inflammatory 3α/CCL20 is expressed at inflamed epithelial cell surfaces and is the most potent chemokine known in attracting Langerhans cell precursors. J Exp Med 2000; 192:705-717.
48. Cavani A, Nasorri F, Prezzi C et al. Human $CD4^+$ T lymphocytes with remarkable regulatory functions on dendritic cells and nickel-specific Th1 immune responses. J Invest Dermatol 2000; 14:395-402.
49. Enk AH, Katz SI. Identification and induction of keratinocyte-derived IL-10. J Immunol 1992; 149:92-95.
50. Teunissen MB, Koomen CW, Jansen J et al. In contrast to their murine counterparts, normal human keratinocytes and human epidermoid cell lines A431 and HaCaT fail to express IL-10 mRNA and protein. Clin Exp Immunol 1997; 107:213-223.
51. Meng X, Sawamura D, Tamai K et al. Keratinocyte gene therapy for systemic diseases. Circulating interleukin 10 released from gene-transferred keratinocytes inhibits contact hypersensitivity at distant areas of the skin. J Clin Invest 1998; 101:1462-1467.
52. Ding L, Linsley PS, Huang LY et al. IL-10 inhibits macrophage costimulatory activity by selectively inhibiting the up-regulation of B7 expression. J Immunol 1993; 151:1224-1234.
53. Michel G, Mirmohammadsadegh A, Olasz E et al. Demonstration and functional analysis of IL-10 receptors in human epidermal cells: decreased expression in psoriatic skin, down-modulation by IL-8, and up-regulation by an antipsoriatic glucocorticosteroid in normal cultured keratinocytes. J Immunol 1997; 159:6291-6297.
54. Seifert M, Sterry W, Effenberger E et al. The antipsoriatic activity of IL-10 is rather caused by effects on peripheral blood cells than by a direct effect on human keratinocytes. Arch Drmatol Res 2000; 292:164-172.
55. Epstein SP, Baer RL, Thorbecke GJ et al. Immunosuppressive effects of transforming growth factor β: inhibition of the induction of Ia antigen on Langerhans cells by cytokines and of the contact hypersensitivity response. J Invest Dermatol 1991; 96:832-837.
56. Muzio M, Sironi M, Polentarutti N et al. Induction of transforming growth factor-β1 of the interlekin-1 receptor antagonist and of its intracellular form in human polymorphonuclear cells. Eur J Immunol 1994; 24:3194-3198.
57. Nickoloff BJ, Turka L. Immunological functions of non-professional antigen-presenting cells: new insights from studies of T-cell interactions with keratinocytes. Immunol Today 1994; 15:464-469.
58. Gaspari AA, Jenkins MK, Katz SI. Class II MHC-bearing keratinocytes induce antigen-specific unresponsiveness in hapten-specific Th1 clones. J Immunol 1988; 141:2216-2220.
59 Gaspari AA, Katz SI. Induction of in vivo hyporesponsiveness to contact allergen by hapten-modified Ia^+ keratinocytes. J Immunol 1991; 147:4155-4161.
60. Federici M, Giustizieri ML, Scarponi C et al. Impaired IFN-γ-dependent inflammatory responses in human keratinocytes overexpressing the suppressor of cytokine signaling 1. J Immunol 2002; 168:434-442.
61. Yasukawa H, Sasaki A, Yoshimura A. Negative regulation of cytokine signalling . Annu Rev Immunol 2000; 18:143-160.

CHAPTER 11

UVB Radiation and Modulation of T Cell Responses to Haptens

Thomas Schwarz, Stefan Beissert and Agatha Schwarz

Summary

Ultraviolet (UV)B radiation exerts a variety of biological effects, including premature skin ageing, induction of skin cancer and suppression of immune responses. The implications of the immunosuppressive properties of UVB radiation are severalfold since UVB-induced immunosuppression of the immune system is not only responsible for the exacerbation of infectious diseases following UVB exposure, but also contributes to the induction of skin cancer. Therefore, detailed knowledge about the mechanisms underlying UVB-mediated immunosuppression is of utmost importance. The majority of the mechanistic studies trying to elucidate the complex phenomena employed the model of sensitization against haptens. Studies utilizing this model provided clear evidence that UVB radiation does not only inhibit the induction of contact hypersensitivity (CHS) but also induces hapten-specific tolerance which is due to the generation of hapten-specific suppressor/regulatory T cells. The model of hapten sensitization did not only provide important information for photoimmunology but also for immunology in general. In the following the impacts of UVB radiation on T cell responses to haptens and its implications for photoimmunology will be briefly reviewed.

UV radiation can be regarded as one of the most significant environmental factors affecting humans. In addition to its essential ecological impacts and to its indispensable effects on life of humans, animals and plants, UV radiation, in particular the middle wave length range (UVB, 290-320 nm), can also exert hazardous effects on health. These include induction of inflammation and cell death, premature skin aging, exacerbation of infectious diseases and induction of skin cancer, the malignancy with the most rapidly increasing incidence world wide.[1-4] When studying the biological effects of UVB radiation it became evident that UVB radiation can also significantly compromise the immune system. Thanks to numerous studies in the field of photoimmunology over the last 30 years, it became much clearer by which pathways UVB radiation suppresses the immune system. The by far most frequently used model to address these issues is the induction of delayed type hypersensitivity (DTH) and CHS to haptens, low molecular weight reactive chemical compounds which become immunogenic upon being bound to carrier proteins. The following chapter will discuss the currents status of the effects of UVB radiation on T cell responses to haptens.

Immune Mechanisms in Allergic Contact Dermatitis, edited by Andrea Cavani and Giampiero Girolomoni. ©2005 Eurekah.com.

UVB Radiation Inhibits the Induction of CHS Against Haptens

Photoimmunology started in the mid seventies with the observation of a link between UVB radiation and immune suppression during photocarcinogenesis. UVB-induced skin tumors were recognized to be highly antigenic since they were rejected upon inoculation into syngeneic healthy recipients.[5-7] However, pharmacologically immunosuppressed recipients were not able to reject the transplanted tumors and the same was observed when the recipients received instead of immunosuppressive drugs treatment with UVB radiation,[8-10] clearly indicating that UVB radiation is able to exert immunosuppressive properties.

Since transplantation of tumors is complicated and labour-intensive researchers were seeking for easier-to-handle models to investigate the mechanisms underlying UVB-induced immunosuppression. Toews et al were the first to use hapten-mediated CHS to study these effects.[11] The vast majority of the following studies used this model. CHS is a special form of a DTH response which is induced by epicutaneous application of haptens, e.g., dinitrofluorobencene, trinitorchlorobencene, oxazolone or fluorescein isothiocyanate (FITC). Painting of haptens to skin which had been irradiated with low doses of UVB (around 1 J/m^2) failed to induce CHS, while application of the hapten at a non-UVB-exposed skin area caused a normal CHS response.[11] The alteration of the immune response by UVB radiation was associated with a reduction in the number of Langerhans cells (LCs), the primary antigen presenting cell within the epidermis.[11] Accordingly, the sensitization was much weaker when the hapten was applied on skin areas which constitutively harbour fewer LCs, e.g., the tail region. The reduction and the changes in the morphology of LCs by UVB radiation were confirmed shortly thereafter by Aberer et al.[12] Since UVB exposure and sensitization affect the same skin area, this model is also referred to as local immunosuppression, since later studies showed that UVB can also affect the immune system in a systemic fashion (see section "UVB Radiation Inhibits Sensitization to Haptens in a Systemic Fashion" below).

A number of in vitro studies have shown that UVB radiation impairs LCs in their activity to present antigens.[13,14] It was also demonstrated that low dose UVB-exposed LCs preferentially activate $CD4^+$ cells of the T helper 2 subset, but do not activate T helper 1 cells.[15] A follow-up study on this reported that UVB radiation converts LCs from immunogenic to tolerogenic antigen-presenting cells because of induction of specific clonal anergy in $CD4^+$ T helper 1 cells.[16] However, one has to be careful when extrapolating these findings to the in vivo situation for hapten sensitization, since these studies used either allogeneic primary systems or primed syngeneic systems, neither one of these being suitable as serving as an appropriate surrogate for the induction of hapten sensitization and its suppression by UVB radiation since hapten sensitization represents a primary syngeneic response.

The capacity of low dose UVB to suppress the induction of CHS seems to be genetically restricted, since inhibition of CHS was only observed in some particular strains (e.g., C3H/HeN, C57BL/6), designated UVB-susceptible, while other strains (C3H/HeJ, Balb/c) reacted with normal sensitization despite UVB exposure, designated UVB-resistant.[17] The phenotypic traits of UVB-resistant and -susceptible mice are polygenically inherited. The relevant loci at which polymorphic alleles reside are *LPS* and *TNFα*.[18] Indeed, tumor necrosis factor α (TNF-α) appears to be involved in the suppression of CHS by low dose UVB radiation, since suppression can be prevented when the UVB-susceptible mice are treated with neutralizing anti-TNF-α antibodies.[19] Accordingly, the induction of CHS responses was significantly reduced when the hapten was painted on murine skin into which subinflammatory doses of TNF-α had been injected.[19] Moreover, it has been recognized that whether mice react in an UVB-resistant or -susceptible fashion also depends on the time lag between UVB exposure and sensitization[20] and on the concentration of the hapten.[21]

The phenotypes of UVB-susceptibility and UVB-resistance were also detectable in humans.[22] In addition, it was shown that the susceptibility to the effects of UVB radiation on induction of CHS may represent a risk factor for skin cancer, since the incidence of UVB susceptibility was significantly higher in skin cancer patients. The same group very recently showed a strong association of UVB-susceptible and UVB-resistant phenotypes in humans with microsatellite markers and single nucleotide polymorphisms in the TNF region, suggesting the TNF region as a good candidate for containing genes that determine UVB susceptibility in humans.[23]

UVB Radiation Induces Hapten-Specific Tolerance

Even more remarkable than the fact that UVB radiation inhibits the induction of CHS was the observation by Toews et al that UVB radiation even induces hapten-specific tolerance.[11] Mice which received the initial hapten painting on UVB-exposed skin could not be resensitized with the same hapten at a later time point.[11] Since the very same mice could be immunized against another unrelated hapten without any problems this indicated that UVB radiation does not suppress the immune system in a general but in an antigen-specific manner. Subsequent investigations showed that hapten-specific tolerance was due to the induction of suppressor/regulatory T cells (see section "UVB-Induced Tolerance Is Mediated by Suppressor/Regulatoty T Cells" below) and occurred also in the model of systemic immunosuppression (see section "UVB Radiation Inhibits Sensitization to Haptens in a Systemic Fashion" below).

UVB radiation induces hapten-specific tolerance also in humans. In one study, only 10% of the human subjects did not develop specific CHS upon subsequent reimmunization with the same hapten.[22] Nevertheless, tolerance induced in these few volunteers was hapten-specific, since they reacted with pronounced CHS responses upon subsequent sensitization with another, non-related hapten.[22] Cooper et al reported that a higher percentage of human volunteers developed tolerance when the hapten was initially painted onto skin areas exposed to erythemogenic UVB doses.[24] These differences may be due to the different protocols. Nevertheless, both reports demonstrate the existence of a subtype of humans who develop tolerance when the hapten is first applied onto UVB-exposed skin.

Cooper et al[24] also showed that erythemogenic UVB doses did not only deplete LCs but also induced the infiltration of $CD1a^-$ $HLA\text{-}DR^+$ $CD36^+$ macrophages. These macrophages were able to activate autoreactive T cells,[25,26] specifically $CD4^+$ suppressor-inducer cells,[27] which in turn induced the maturation of suppressor T cells.[28] Furthermore, these macrophages which also express $CD11b^+$ were identified to potently release the immunosuppressive cytokine interleukin (IL)-10, representing the major source for epidermal IL-10 protein in the human system.[29] This finding appears to be of relevance since IL-10 seems to be involved in UVB-induced immunosuppression (see section "UVB Radiation Inhibits Sensitization to Haptens in a Systemic Fashion" below). In addition, these macrophages infiltrating the epidermis upon UVB exposure induced in vitro $CD4^+$ T lymphocytes which lack the expression of the IL-2 receptor alpha chain.[30] Downregulation of the IL-2 receptor alpha chain seems to be caused by transforming growth factor β, another immunosuppressive mediator.

UVB-Induced Tolerance Is Mediated by Suppressor/Regulatory T Cells

Suppression of tumor rejection by UVB radiation could be transferred by injecting T lymphocytes from UVB-irradiated mice into normal recipients, which subsequently became unable to reject UVB-induced tumors.[31,32] The same was observed in the model of hapten sensitization.[33,34] UVB-mediated suppression of CHS could be transferred by injecting T lymphocytes obtained from lymph nodes and spleens, respectively, of UVB-exposed and hapten-treated mice into naive recipients. This suppression is hapten-specific, since though not responding to the specific hapten the recipients generated a normal CHS response against a non-cross reacting hapten.[33,34] Together, these findings suggested that tolerance is mediated via induction of hapten-specific T suppressor cells. However, the phenotype of these postulated cells and the molecular mechanisms underlying these suppressive phenomena were only poorly characterized. Consequently, the term T suppressor cells was almost banned and the entire concept of suppression drawn into question. Nevertheless, the concept of T suppressor cells was persistently pursued in the field of photoimmunology. The final discovery of regulatory T cells (see below) revealed that this persistence was justified and right.[35]

Transfer of tolerance could be observed in both the local and systemic immunosuppression. However, different types of T cells seem to be responsible for the transfer of suppression. In the systemic form of UVB-induced suppression (see section "UVB Radiation Inhibits Sensitization to Haptens in a Systemic Fashion" below) transfer of tolerance is mediated by the induction of antigen-specific, $CD3^+$, $CD4^+$, and $CD8^-$ suppressor cells.[36] Using the local form of UVB-induced immunosuppression, Elmets et al[33] demonstrated that treatment of cells from UVB-irradiated animals with antibodies directed against Lyt-1 (CD5) completely abrogated their ability to transfer suppression, while treatment of cells with antibodies directed against Lyt-2 (CD8) inhibited suppression partially.[33] Accordingly, Schwarz et al reported that in the low dose model transfer of suppression was lost when T lymphocytes were depleted of $CD8^+$ cells.[37] It is important to note that T suppressor cells can only affect the induction but not the elicitation of CHS, since transfer of UVB-induced T suppressor cells into already sensitized mice does not affect the CHS response in the recipients.[38] This indicates that effector T cells dominate suppressor T cells.

The field of suppression and suppressor T cells experienced resuscitation by the discovery of regulatory T cells. Chronic activation of both human and murine $CD4^+$ T cells in the presence of IL-10 induced $CD4^+$ T cell clones with low proliferative capacity, producing high levels of IL-10, low levels of IL-2 and no IL-4.[39] These antigen-specific T cell clones suppressed the proliferation of $CD4^+$ T cells in response to antigen, and prevented T cell-mediated colitis in SCID mice. This $CD4^+$ T-cell subset was designated T regulatory cells. Another subset of $CD4^+$ regulatory T cells is characterized by the constitutive expression of the α chain of the IL-2 receptor (CD25).[40] Interestingly, $CD4^+CD25^+$ regulatory T cells constitute approximately 5-10% of all murine peripheral $CD4^+$ T cells. These new developments inspired many studies investigating the role of suppressor/regulatory T cells, currently one of the most intensively studied subjects in general immunology. Whether the cells are called regulatory or suppressor represents more a semantic issue. In any case, because of these new discoveries T suppressor cells are now "socially accepted" in the immunologic community.[41]

Shreedhar et al were the first to clone T cells from UVB-exposed mice which were subsequently sensitized with FITC. Cells cloned from UVB-exposed mice were $CD4^+$, $CD8^-$, TCR-α/β^+, MHC restricted and specific for FITC. They produced IL-10, but not IL-4 or interferon-γ, whereas cells from unirradiated animals produced high amounts of interferon-γ and little IL-4 and IL-10.[42] The cytokine pattern of the UVB-induced cells was related but not identical to that of T regulatory 1 cells, thus the authors designated these cells to the T regulatory 2 type. In vitro these cells blocked antigen presenting cell functions and IL-12 production. Even more importantly, injection of these T cells into untreated recipients suppressed the induction of CHS against FITC.

While previously suppressor T cells were mostly allocated to the CD8 type there is increasing evidence that the majority belong to the CD4 type. In this respect, the role of $CD4^+CD25^+$ regulatory T cells in mediating UVB-induced tolerance remains to be determined. The importance of $CD4^+$ T cells in mediating UVB-induced immunosuppression was recently confirmed using MHC class II knock-out mice. The animals turned out to be resistant to the immunosuppressive effects of UVB radiation, indicating that UVB-induced immunosuppression is due to preferential activation of $CD4^+$ suppressor/regulatory T cells and not due to deficient priming and expansion of effector $CD8^+$ T cells.[43]

There is evidence that T suppressor cells express the negative regulatory molecule cytotoxic T lymphocyte activation molecule-4 (CTLA-4) on their surface. CTLA-4 appears to be functionally relevant since inhibition of CTLA-4 by a neutralizing antibody inhibits tolerance and transfer of suppression.[44] In vitro stimulation of T suppressor cells induces the release of IL-2, interferon-γ, high amounts of IL-10 but no IL-4, a cytokine secretion pattern reminiscent of that of regulatory T cells. Release of IL-10 induced by CTLA-4 activation appears to be functionally relevant since transfer of suppression could be inhibited when recipients received neutralizing anti-IL-10-antibodies.

The distinctive heterogeneity of suppressor cells became even more evident by the observation that UVB-induced natural killer (NK) T cells are involved in the suppression of tumor immune responses.[45] NKT cells express intermediate amounts of T cell receptor molecules and coexpress surface antigens normally found on natural killer cells (NK1.1, DX5 and Ly49a). Moodycliffe et al provided convincing evidence that UVB-induced suppressor T cells belong to the NKT type and suppress DTH and tumor immunity. Because of the latter observation these cells may play a critical role in regulating the growth of UVB-induced skin cancers.

UVB Radiation Inhibits Sensitization to Haptens in a Systemic Fashion

When applying higher UVB doses (around 2000 J/m^2) it was observed that CHS cannot be induced in mice even if the hapten was applied at unirradiated sites.[46] This experiment clearly indicated that UVB radiation may suppress the immune system also in a systemic fashion (systemic immunosuppression). As it was the case for the local model (see section "UVB Radiation Inhibits the Induction of CHS Against Haptens" above), systemic immunosuppression did not only inhibit the induction of CHS against haptens but also induced hapten-specific tolerance.[46]

The mechanism by which UVB radiation inhibits an immune response to a hapten applied at a distant non-UVB-exposed skin area remained unclear for many years.[34,46] However, it was obvious that different mechanisms than in the local model should be responsible since the LCs critically involved in local immunosuppression[11] were not altered both in their number and their morphology in non-UVB-exposed skin areas.[47] Since keratinocytes were identified to be a potent source for cytokines[48] and by virtue of their anatomic location are the natural target for UVB radiation, Schwarz et al proposed that UVB-exposed keratinocytes could be the source for immunosuppressive soluble mediators which enter the circulation and thereby may exert systemic immunosuppressive effects.[49] Accordingly, it was demonstrated that intravenous injection of supernatants obtained from UVB-irradiated murine keratinocytes into naive mice before hapten application resulted in suppression of the induction of CHS.[49] This observation opened the field of UVB-induced immunosuppressive soluble mediators with many important follow-up studies.

The concept of UVB-exposed keratinocytes as a source of immunosuppressive factors has been supported by a variety of other studies. For example, injection of supernatants from UVB-exposed murine keratinocytes suppressed the induction of DTH to alloantigen and trinitrophenyl-modified self antigens in syngeneic and allogeneic mice.[50]

With regard to immunosuppresssion two cytokines appear of special interest, i.e., TNF-α and IL-10. While TNF-α seem to be primarily involved in local immunosuppression (see section "UVB Radiation Inhibits the Induction of CHS Against Haptens" above), IL-10 is certainly the major player in systemic immunosuppression. IL-10 was found to be able to interfere with the antigen-presenting capacity of LCs. In vitro incubation of LCs with IL-10 abrogated the ability of these cells to present antigen to Th1 clones and even tolerized them.[51] Besides macrophages, B and T cells, keratinocytes can function as a source for IL-10.[52,53] IL-10 production in the skin is enhanced upon application of contact allergens whereas tolerogens or irritants have no effect.[54] In addition, injection of IL-10 into the skin area of hapten application prevents the induction of CHS and induces hapten specific tolerance.[55] In addition, it was demonstrated that UVB radiation upregulates IL-10 in both murine and human keratinocytes.[53] Due to its ability to downregulate inflammatory and immune reactions, keratinocyte-derived IL-10 may play an important role in UVB-induced immunosuppression.[53] Accordingly, injection of a neutralizing anti-IL-10-antibody into UVB-irradiated mice prevented systemic UVB-induced suppression of the induction of DTH.[53] Moreover, it was observed that spleen cells from UVB-exposed animals do not present antigen to T helper 1 cells, while presentation to T helper 2 cells was even enhanced.[56] Both of these effects could be prevented by injection of an anti-IL-10 antibody, suggesting a crucial role of IL-10 in UVB-induced systemic immunosuppression. On the other hand, IL-10 was also found to influence not only the induction, but also the elicitation phase of both CHS and DTH.[57] As a whole, these data suggest that UVB radiation with the help of IL-10 tolerizes T helper 1 cells and activates T helper 2 cells.

The T helper 2 shift in systemic immunosuppression is further supported by the observation that immune suppression is blocked in mice treated with an anti-IL-4 antiserum.[58] However, UVB radiation does not directly induce IL-4 release. This might be mediated indirectly via the UVB-induced release of prostaglandin E_2 by keratinocytes. Accordingly, cylcooxygenase-2 inhibitors blocked IL-4 production. This suggests that UVB exposure activates a cytokine cascade (prostaglandin $E_2 \rightarrow$ IL-4 $\rightarrow$ IL-10) that finally results in systemic immunosuppression.[58] In the human system, it was recently observed that UVB radiation stimulates the immigration of neutrophils which via secretion of IL-4 might favour type 2 T cell responses in UVB-exposed skin.[59] Hence, there is profound evidence that in vivo exposure to UVB radiation induces a shift towards a Th2 immune response in vivo, explaining that mostly T helper 1 mediated cellular immune reactions are impaired by UVB radiation.

IL-10, however, does not only appear to be involved in mediating tolerance induced by high dose UVB, but also in tolerance induced by low dose UVB. Niizeki and Streilein observed that (i) tolerance develops when haptens are applied onto skin areas in which IL-10 was injected and (ii) that intraperitoneal injection of an anti-IL10 antibody prevented tolerance induced by low dose UVB.[60]

Cis-Urocanic Acid

Cis-urocanic acid (cis-UCA) is the photoisomer of trans-UCA, a natural component of the stratum corneum.[61] Removal of the stratum corneum by tape stripping prevented UVB-induced inhibition of hapten sensitization,[62] thus it was supposed that cis-UCA is involved in UVB-induced immunosuppression. In fact, cis-UCA exerts immunosuppressive properties, since hapten application onto skin in which cis-UCA was injected did not result in sensitization but resulted in hapten specific tolerance which could be even adoptively transferred.[63] Although cis-UCA is certainly involved in UVB-induced immunosuppression the mechanisms involved in UVB- and cis-UCA-mediated immunosuppression do not appear to be exactly the same. For example, anti-IL-10-antibodies blocked the induction of UVB-induced hapten-specific tolerance completely, while cis-UCA induced tolerance only partially.[60] In addition, complete inhibition of UVB-induced tolerance by application of antibodies against cis-UCA,[64] has not been demonstrated yet. cis-UCA was recently shown to induce the release of neuropeptides[65] which might be involved in UVB-induced immunosuppression (see section "Mast Cells" below).

Neuropeptides

The demonstration of connections between nerve fibers, cutaneous and immune cells and the observation that all these cells exhibit the capacity to release neurotransmitters, in particular neuropeptides, has led to the concept of the neuro-immuno-cutaneous system.[66] Not surprisingly, these mediators appear to be involved in UVB-induced immunosuppression as well.

Proopiomelanocortin-derived peptides like α-melanocyte stimulating hormone (α-MSH) and ACTH are produced by keratinocytes. Their release is induced by UVB radiation. Injection of α-MSH into mice prevented the induction of CHS and induced hapten-specific tolerance.[67] This effect appeared to be mediated via IL-10 since this effect could be prevented by injection of neutralizing anti-IL-10-antibodies. Although it appears to be quite likely that α-MSH is involved in mediating UVB-induced immunosuppression, functional studies e.g., using melanocortin receptor-deficient mice are still missing.

Calcitonin gene-related peptide (CGRP) has been recognized to modulate the function of LCs through closely associated CGRP positive nerve fibers.[68] UVB radiation induces the release of CGRP from cutaneous nerve endings.[69,70] CGRP seems to be involved in the inhibition of CHS by low dose UVB radiation via triggering the release of TNF-α from dermal mast cells.[70] The functional role of CGRP in inhibiting induction of CHS following UVB exposure was proven by the observation that CGRP antagonists restored the hapten sensitization in UVB-exposed skin.[70,71] CGRP also participates in promoting hapten-specific tolerance after UVB exposure via inducing IL-10. In contrast to TNF-α, IL-10 does not appear to be released from mast cells since CGRP induces hapten-specific tolerance also in the absence of mast cells.[72]

Mast Cells

For a long time mast cells (MCs) have been almost completely ignored in photoimmunology. Several recent studies, however, have demonstrated the crucial role MCs might play in UVB-induced immunosuppression at least in the murine system. Hart et al found that the prevalence of histamine-staining dermal MCs in different strains of mice correlated directly with their susceptibility to UVB-induced systemic immunosuppression.[73] MC-deficient (*Wf/Wf*) mice were resistant to systemic UVB-induced immunosuppression. Reconstitution of MC-deficient mice with bone marrow-derived MC precursors from wild-type mice rendered the mice susceptible to UVB irradiation for systemic suppression of CHS responses. In contrast, UVB radiation suppressed CHS responses in MC-deficient mice when the hapten was painted onto the UVB-exposed skin area. In addition, a correlation between the prevalence of dermal MC and the degree of susceptibility of different strains of mice to the immunomodulatory effects of cis-UCA was observed.[74] This appears to be functionally relevant since MC-deficient mice were rendered susceptible to cis-UCA-mediated immunosuppression only after reconstitution of the skin with bone marrow-derived MCs, suggesting that MCs are mediating cis-UCA-induced immunomodulation.

However, there is also evidence from the Streilein group that MCs can be involved in local UVB-induced immunosuppression. Upon triggering histamine release from MCs with antibodies they observed that induction of CHS was impaired in UVB-susceptible mice (C3H/HeN) but not in UVB-resistant (C3H/HeJ), and MC-deficient mice (*Sl/Sl^d*). In addition, local suppression of CHS by UVB radiation was not impaired in MC-deficient mice.[75] Since injection of neutralizing antibodies against TNF-α inhibited UVB-induced suppression in UVB-susceptible mice, the authors concluded that MC-derived TNF-α might interfere with hapten-specific signals inducing a CHS response. There is recent evidence from the same group that MCs may also be a source for IL-10 which mediates tolerance in UVB-induced local immunosuppression.[76]

UVB-Induced Inhibition of CHS Induction and UVB-Induced Hapten-Specific Tolerance Are Mediated Via Different Mechanisms

Since impairment of CHS and induction of tolerance are the consequence of the same event, i.e., low or high dose UVB exposure followed by hapten application, it was thought for years that the same mechanisms are involved. Irradiation of skin either with low or high dose UVB results in depletion of epidermal LCs[12] which are crucial for the induction of CHS.[77] Under normal conditions LCs take up the hapten, migrate with the hapten to the regional lymph nodes where the LCs, which have matured during their journey into potent immunostimulatory dendritic cells, present the antigen to T lymphocytes.[78] Thus, impairment of the induction of CHS upon application of the hapten onto UVB-exposed skin can be easily explained by the depletion/alteration of LCs. However, this does not apply for systemic UVB-mediated immunosuppression, since even after multiple UVB-exposures LCs are not affected in skin areas which are not directly exposed to UVB radiation.[47] In this case, secretion of soluble mediators by keratinocytes appears to be of importance (see section "UVB Radiation Inhibits Sensitization to Haptens in a Systemic Fashion" above).

Since application of haptens onto UVB-exposed mice results both in the impairment of CHS and induction of hapten specific tolerance it was postulated for quite a long time that the same mechanisms are involved and that induction of tolerance is the consequence of the inhibition of CHS induction. However, there is accumulating evidence that the molecular basis of UVB-induced tolerance is different from the mechanism responsible for UVB-impaired induction of CHS. Although TNF-α induced similar alterations in epidermal LCs like UVB[79] and although injection of anti-TNF-α antibodies inhibited UVB-mediated impairment of CHS, development of tolerance could not be prevented by injection of neutralizing TNF-α antibodies.[63,80]

Hammerberg et al observed that application of a hapten immediately after a single low dose UVB exposure resulted in inhibition of the induction of CHS but failed to induce tolerance.[81] A state of tolerance could only be achieved if a delay of 72 hours was allowed to elapse between the UVB exposure and the initial sensitization. 72 hours after in vivo UVB exposure Langerhans cells were depleted, and Ia^+CD11b^{bright} macrophages had appeared in the epidermis. Intracutaneous injection of haptenated epidermal cell suspensions obtained from in vivo UVB-exposed skin induced hapten-specific tolerance. Induction of tolerance was lost when $CD11b^{bright}$ cells were removed from the epidermal cell suspensions before injection. In addition, induction of UVB-mediated tolerance was blocked when mice were treated with an antibody blocking CD11b.[82] CD11b can serve as a receptor for the fragment of the complement component 3, iC3b. To test whether C3 activation is critically required for UVB-induced immunosuppression, mice with a genetic disruption of the C3 gene were used. Indeed, C3-deficient mice did not develop tolerance following hapten application through a skin area which received a single low dose UVB exposure.[83] Inhibition of C3 activation partially blocked UVB-induced infiltration of $CD11b^{bright}$ macrophages but did not prevent UVB-induced depletion of Ia^+CD11b^{low} LCs. Together, these findings indicate that the ability of UVB-exposed skin to induce tolerance is critically dependent on inflammatory Ia^+CD11b^{bright} monocytic/macrophagic cells, which infiltrate UVB-exposed skin, and not on the depletion of LCs.

IL-12 Prevents UVB-Induced Immunosuppression and Breaks Hapten-Specific Tolerance

IL-12 is a heterodimeric cytokine (p70) consisting of two chains (p35, p40). Among its many biological activities it supports the development of a Th1 immune response.[84] The crucial functional role of IL-12 during epicutaneous hapten sensitization was demonstrated by the observation that intraperitoneal injection of a neutralizing anti-IL-12-antibody into naive mice before epicutaneous painting of haptens resulted in the inability to induce sensitization in these animals.[85,86] Furthermore, blocking of IL-12 before sensitization induced hapten-specific tolerance, since mice initially anti-IL-12-treated and sensitized could not be resensitized with the same hapten.[85]

IL-12 and IL-10 represent a kind of immunological opponents.[87] Thus, it was obvious to study whether IL-12 can overcome systemic UVB-induced immunosuppression. Indeed, injection of IL-12 blocked systemic suppression of both CHS and DTH.[88] Since adoptive transfer of spleen cells from UVB-exposed animals treated with IL-12 had no effect on the CHS response of the recipient mice, it was concluded that IL-12 may prevent the generation of UVB-induced T suppressor cells.[88]

However, IL-12 is also able to prevent local immunosuppression induced by UVB radiation. A single intraperitoneal injection of IL-12 administered between UVB exposure and hapten application to the irradiated skin area completely restored the CHS response.[89] Moreover, IL-12 also breaks UVB-induced tolerance since UVB-tolerized mice could be successfully sensititzed with the specific hapten when the animals had received IL-12 intraperitoneally.[89] Recently, IL-12 was also shown to be able to revert the immunosuppressive effects of cis-UCA.[90]

The mechanisms by which IL-12 prevents UVB-induced immunosuppression and in particular breaks established tolerance still remain to be determined. Ando et al reported that T cells obtained from mice which were immunosuppressed by UVB radiation in a systemic fashion were impaired in their capacity to release interferon-γ despite the presence of the specific antigen and antigen-presenting cells.[91] Addition of IL-12 restored the defective interferon-γ production. An alternative explanation might be the capacity of IL-12 to rescue antigen presenting cells from apoptosis induced by T suppressor cells (see section "The Role of Apoptosis in Hapten-Specific UVB-Induced Tolerance" below).

The Role of Apoptosis in Hapten-Specific UVB-Induced Tolerance

Although the involvement of T suppressor cells in UVB-induced tolerance has been described two decades ago,[33] the mechanisms by which these cells mediate suppression still remain unclear. However, evidence exists that apoptosis and the death receptor system Fas/CD95 may be involved. It was observed that mice deficient in either the Fas receptor (*lpr*) or the Fas ligand (*gld*) responded to UVB radiation with an impaired CHS response. However, in contrast to UVB-exposed wild type mice, UVB-irradiated *lpr* and *gld* mice did not develop tolerance, clearly suggesting that the Fas/FasL system is crucial for the development of tolerance.[92] In vitro studies implied that T cells from UVB-tolerized mice might drive antigen-presenting cells into apoptosis in the presence of the specific hapten. Accordingly, IL-12 could rescue the majority of dendritic cells under these conditions.

The critical role of the Fas/Fas-ligand interactions in UVB-induced immunosuppression was also demonstrated in another study, using the high dose UVB model. This study demonstrated that donor- but not recipient-derived FasL expression was critically required for both the generation and the function of UVB-induced suppressor T cells.[93] These findings differ from the study mentioned above[92] which may reflect differences when studying local (low dose) or systemic (high dose) UVB-induced immunosuppression. In any case, apoptosis and the Fas/FasL system appear to be essential for both types of suppression.

Molecular Cellular Targets Involved in UVB-Induced Immunosuppression

Nuclear DNA is the major UVB-absorbing cellular chromophore. UVB radiation induces primarily two types of photolesions in DNA, cyclobutane pyrimidine dimers (CPD) and

<6-4>-photoproducts. For a long time UVB-induced DNA damage had been proposed to play an essential role in UVB-induced immunosuppression. Data to support this hypothesis originated from the marsupial model *Monodelphis domestica*. Unlike humans this animal exhibits the capacity to remove UVB-induced DNA lesions via a repair process called photoreactivation.[94] Accordingly, DNA damage is much faster removed when these animals are exposed to visible light after UVB radiation. UVB-mediated inhibition of the induction of CHS was significantly reduced upon exposure of the animals to visible light immediately after UVB radiation. Since this process removes DNA lesions it was concluded that UVB-induced DNA damage is critically involved in signalling UVB-induced immunosuppression.

This hypothesis was also confirmed by studies using the DNA excision repair enzyme T4 endonuclease V (T4N5) which accelerates removal of DNA lesions.[95] Topical application of T4N5 incorporated into liposomes onto the UVB-exposed skin area significantly antagonized the inhibition of the induction of CHS.[96] There is also evidence that DNA damage triggers cytokine release because UVB-induced production of IL-10 and TNF-α by keratinocytes was significantly suppressed after addition of T4N5.[97,98] In addition, mice in which particular components of the nucleotide excision repair have been knocked out are more susceptible to UVB-induced immunosuppression. With this model it was shown that both global and transcription-coupled repair are needed to mitigate immunomodulation by UVB.[99] The important role of DNA damage in UVB-induced immunosuppression was recently also confirmed in humans in vivo. Utilizing volunteers hypersensitive to nickel, Stege et al showed that UVB-induced suppression of nickel specific hypersensitivity reactions could be prevented when photolyase, a DNA damage repair enzyme, was applied immediately after UVB exposure.[100]

A very recent publication identified a potential link between DNA damage and IL-12.[101] Surprisingly, IL-12 was found to be able both in vitro and in vivo to reduce UVB-induced DNA damage both in mice and men. Reduction of cyclobutane pyrimidine dimers appeared to be mediated via nucleotide excision repair, the endogenous repair system, since this unique effect of IL-12 was not observed in mice in which the nucleotide excision repair was knocked out. Considering the fact that IL-12 prevents UVB-induced immunosuppression and that UVB-induced DNA damage is the primary mediator of UVB-induced immunosuppression, it is tempting to speculate whether at least part of the immunoreconstitutive effects of IL-12 are due to its ability to reduce DNA damage. The observation that IL-12 inhibits UVB-induced IL-10 release[102] which is clearly mediated via DNA damage[97] nourishes this speculation.

Although, there is strong evidence that DNA damage is the crucial molecular mediator of UVB-induced immunosuppression, there is also evidence for extranuclear cellular UVB targets involved in photoimmunology. It was discovered that UVB radiation does not only affect the release of cytokines but may also interfere with the biological activities of these mediators. Accordingly, it was shown that UVB radiation interferes with the signal transduction pathway of interferon-γ and IL-2, both important immunomodulatory cytokines.[103,104] UVB radiation may exert this effect by inhibiting the phosphorylation of important signal transduction proteins involved in the signalling of these two cytokines.

Conclusion

UVB radiation can compromise the immune system very effectively and in several ways. Numerous studies in photoimmunology have contributed to a better understanding how UVB radiation suppresses the immune system and subsequently to a better understanding of the biological and pathological effects of UVB radiation. In addition, a variety of photoimmunological studies yielded important information not only for photoimmunology but for immunology in general. In this respect, the usage of the model of hapten sensitization was a very useful tool. On the other hand, the in vivo situation when dealing with solar radiation may even be more complex since the solar spectrum in addition to UVB also contains the long wave length range UVA (320-400 nm). Although less intensively studied, there is clear evidence that UVA radiation influences the immune system as well.[105,106] In addition, it may interfere with the effects of UVB. How these two spectra interact with each other will be demonstrated in the future.[107] In any case, the model of hapten sensitization will be of great benefit for these purposes.

Acknowledgments

T.S is supported by grants from the German Research Foundation (DFG, SFB 293, B9) and the European Community (QLK4-CT-2001-00115). S.B. is supported by a grant from the German Research Foundation (DFG, BE 1580/2-3). The authors apologize for not discussing and citing many important papers because of the limitations of space and of the number of references.

References

1. Fisher GJ, Wang ZQ, Datta SC et al. Pathopyhsiology of premature skin aging induced by ultraviolet light. N Engl J Med 1997; 373:1419-1428.
2. Wlaschek M, Tantcheva-Poor I, Naderi L et al. Solar UV irradiation and dermal photoaging. J Photochem Photobiol B 2001; 63:41-51.
3. Kraemer KH. Sunlight and skin cancer: Another link revealed. Proc Natl Acad Sci USA 1997; 94:11-14.
4. Murphy G, Young AR, Wulf HC et al. The molecular determinants of sunburn cell formation. Exp Dermatol 2001; 10:155-160.
5. Kripke ML. Antigenicity of murine skin tumors induced by ultraviolet light. J Natl Cancer Inst 1974; 53:1333-1336.
6. Kripke ML. Latency, histology and antigenicity of tumors induced by ultraviolet light in three inbred mouse strains. Cancer Res 1977; 37:1395-1400.
7. Kripke ML. Effects of UV radiation on tumor immunity. J Natl Cancer Inst 1990; 82:1392-1396.
8. Kripke ML, Lofgreen JS, Beard J et al. In vivo immune responses of mice during carcinogenesis by ultraviolet irradiation. J Natl Cancer Inst 1977; 59:1227-1230.
9. Fisher MS, Kripke ML. Systemic alteration induced in mice by ultraviolet light irradiation and its relationship to ultraviolet carcinogeneis. Proc Natl Acad Sci USA 1977; 74:1688-1692.
10. DeFabo EC, Kripke ML. Dose-response characteristics of immunologic unresponsiveness to UV-induced tumors produced by UV irradiation of mice. Photochem Photobiol 1979; 30:385-390.
11. Toews GB, Bergstresser PR, Streilein JW. Epidermal Langerhans cell density determines whether contact hypersensitivity or unresponsiveness follows skin painting with DNFB. J Immunol 1980; 124:445-453.
12. Aberer W, Schuler G, Stingl G et al. Ultraviolet light depletes surface markers of Langerhans cells. J Invest Dermatol 1981; 76:202-210.
13. Stingl G, Gazze-Stingl LA, Aberer W et al. Antigen presentation by murine epidermal langerhans cells and its alteration by ultraviolet B light. J Immunol 1981; 127:1707-1713.
14. Aberer W, Stingl G, Stingl-Gazze LA et al. Langerhans cells as stimulator cells in the murine primary epidermal cell-lymphocyte reaction: Alteration by UV-B irradiation. J Invest Dermatol 1982; 79:129-135.
15. Simon JC, Cruz Jr PD, Bergstresser PR et al. Low dose ultraviolet B-irradiated Langerhans cells preferentially activate CD4+ cells of the T helper 2 subset. J Immunol 1990; 145:2087-2091.
16. Simon JC, Tigelaar RE, Bergstresser PR et al. Ultraviolet B radiation converts langerhans cells from immunogenic to tolerogenic antigen-presenting cells. Induction of specific clonal anergy in CD4+ T helper 1 cells. J Immunol 1991; 146(2):485-489.
17. Streilein JW, Bergstresser PR. Genetic basis of ultraviolet-B effects on contact hypersensitivity. Immunogenetics 1988; 27:252-258.
18. Streilein JW, Taylor JR, Vincek V et al. Immune surveillance and sunlight-induced skin cancer. Immunol Today 1994; 15:174-179.

19. Yoshikawa T, Streilein JW. Genetic basis of the effects of ultraviolet light B on cutaneous immunity. Evidence that polymorphism at the TNF-α and Lps loci governs susceptibility. Immunogenetics 1990; 32:398-405.
20. Noonan FP, DeFabo EC. Ultraviolet-B dose-response curves for local and systemic immunosuppression are identical. Photochem Photobiol 1990; 52:801-808.
21. Yamawaki M, Katiyar SK, Anderson CY et al. Genetic variation in low-dose UV-induced suppression of contact hypersensitivity and in the skin photocarcinogenesis response. J Invest Dermatol 1998; 111:706-708.
22. Yoshikawa T, Rae V, Bruins-Slot W et al. Susceptibility to effects of UVB radiation on induction of contact hypersensitivity as a risk factor for skin cancer in humans. J Invest Dermatol 1990; 95:530-536.
23. Niizeki H, Naruse T, Hecker KH et al. Polymorphisms in the tumor necrosis factor (TNF) genes are associated with susceptibility to effects of ultraviolet-B radiation on induction of contact hypersensitivity. Tissue Antigens 2001; 58:369-378.
24. Cooper KD, Oberhelman L, Hamilton TA et al. UV exposure reduces immunization rates and promotes tolerance to epicutaneous antigens in humans: Relationship to dose, $CD1a^-DR^+$epidermal macrophage induction, and Langerhans cell depletion. Proc Natl Acad Sci USA 1992; 89:8497-8501.
25. Cooper KD, Fox P, Neises G et al. Effects of ultraviolet radiation on human epidermal cell alloantigen presentation: Initial depression of Langerhans cell-dependent function is followed by the appearance of T6-Dr+ cells that enhance epidermal alloantigen presentation. J Immunol 1985; 134:129-137.
26. Cooper KD, Neises GR, Katz SI. Antigen-presenting $OKM5^+$ melanophages appear in human epidermis after ultraviolet radiation. J Invest Dermatol 1986; 86:363-370.
27. Baadsgaard O, Fox DA, Cooper KD. Human epidermal cells from ultraviolet light-exposed skin preferentially activate autoreactive $CD4^+2H4^+$ suppressor-inducer lymphocytes and $CD8^+$ suppressor/cytotoxic lymphocytes. J Immunol 1988; 140:1738-1744.
28. Baadsgaard O, Salvo B, Mannie A et al. In vivo ultraviolet-exposed human epidermal cells activate T suppressor cell pathways that involve $CD4^+CD45RA^+$ suppressor-inducer T cells. J Immunol 1990; 145:2854-2861.
29. Kang K, Hammerberg C, Meunier L et al. $CD11b^+$ macrophages that infiltrate human epidermis after in vivo ultraviolet exposure potently produce IL-10 and represent the major secretory source of epidermal IL-10 protein. J Immunol 1994; 153:5256-5264.
30. Stevens SR, Shibaki A, Meunier L et al. Suppressor T cell-activating macrophages in ultraviolet-irradiated human skin induce a novel, TGF-beta-dependent form of T cell activation characterized by deficient IL-2r alpha expression. J Immunol 1995; 155:5601-5607.
31. Spellman CW, Woodward JG, Daynes RA. Modification of immunological potential by ultraviolet radiation. Ist ed. Immune status of short-term UV-irradiated mice. Transplantation 1977; 24:112-119.
32. Spellman CW, Daynes RA. Modification of immunological potential by ultraviolet radiation. IInd ed. Generation of suppressor cells in short-term UV-irradiated mice. Transplantation 1977; 24:120-126.
33. Elmets CA, Bergstresser PR, Tigelaar RE et al. Analysis of the mechanism of unresponsiveness produced by haptens painted on skin exposed to low dose ultraviolet radiation. J Exp Med 1983; 158:781-794.
34. Noonan FP, DeFabo EC, Kripke ML. Suppression of contact hypersensitivity by UV radiation and its relationship to UV-induced suppression of tumor immunity. Photochem Photobiol 1981; 34:683-689.
35. Schwarz T. Immunology. In: Bolognia JL, Jorizzo JL, Rapini RP, eds. Dermatology. Mosby in press, 2002.
36. Ullrich SE, McIntryre WB, Rivas JM. Suppression of the immune response to alloantigen by factors released from ultraviolet-irradiated keratinocytes. J Immunol 1990; 145:489-498.
37. Schwarz A, Grabbe St, Mahnke K et al. IL-12 breaks UV light induced immunosuppression by affecting $CD8^+$ rather than $CD4^+$ T cells. J Invest Dermatol 1998; 110:272-276.
38. Glass MJ, Bergstresser PR, Tigelaar RE et al. UVB radiation and DNFB skin painting induce suppressor cells universally in mice. J Invest Dermatol 1990; 94:273-278.

39. Groux H, O'Garra A, Bigler M et al. A CD4+ T-cell subset inhibits antigen-specific T-cell responses and prevents colitis. Nature 1997; 389:737-742.
40. Shevach EM, McHugh RS, Piccirillo CA et al. Control of T-cell activation by CD4+ CD25+ suppressor T cells. Immunol Rev 2001; 182:58-67.
41. Chatenoud L, Salomon B, Bluestone JA. Suppressor T cells—they're back and critical for regulation of autoimmunity! Immunol Rev 2001; 182:149-163.
42. Shreedhar VK, Pride MW, Sun Y et al. Origin and characteristics of ultraviolet-B radiation-induced suppressor T lymphocytes. J Immunol 1998; 161:1327-1335.
43. Krasteva M, Aubin F, Laventurier S et al. MHC class II-KO mice are resistant to the immunosuppressive effects of UV light. Eur J Dermatol 2002; 12:10-19.
44. Schwarz A, Beissert S, Grosse-Heitmeyer K et al. Evidence for functional relevance of CTLA-4 in ultraviolet-radiation-induced tolerance. J Immunol 2000; 165:1824-1831.
45. Moodycliffe AM, Nghiem D, Clydesdale G et al. Immune suppression and skin cancer development: Regulation by NKT cells. Nat Immunol 2000; 1:521-525.
46. Noonan FP, DeFabo EC, Kripke ML. Suppression of contact hypersensitivity by UV radiation: An experimental model. Springer Semin Immunopathol 1981; 4:293-304.
47. Morison WL, Bucana C, Kripke ML. Systemic suppression of contact hypersensitivity by UVB radiation is unrelated to the UVB-induced alterations in the morphology and number of Langerhans cells. Immunology 1984; 52:299-306.
48. Schwarz T, Urbanski A, Luger TA. Ultraviolet light and epidermal cell derived cytokines. In: Luger TA, Schwarz T, eds. Epidermal growth factors and cytokines. New York: Marcel Dekker, 1994; 303-363.
49. Schwarz T, Urbanska A, Gschnait F et al. Inhibition of the induction of contact hypersensitivity by a UV-mediated epidermal cytokine. J Invest Dermatol 1986; 87:289-291.
50. Kim TY, Kripke ML, Ullrich SE. Immunosuppression by factors released from UV-irradiated epidermal cells: Selective effects on the generation of contact and delayed hypersensitivity after exposure to UVA or UVB radiation. J Invest Dermatol 1990; 94:26-32.
51. Enk AH, Angeloni VL, Udey MC et al. Inhibition of Langerhans' cell antigen-presenting function by IL-10. A role for IL-10 in tolerance induction. J Immunol 1993; 151:2390-2398.
52. Enk AH, Katz SI. Identification and induction of keratinocyte-derived IL-10. J Immunol 1992; 149:92-95.
53. Rivas JM, Ullrich SE. Systemic suppression of DTH by supernatants from UV-irradiated keratinocytes: An essential role for IL-10. J Immunol 1992; 148:3133-3139.
54. Enk AH, Katz SI. Early molecular events in the induction phase of contact sensitivity. Proc Natl Acad Sci USA 1992; 89:1398-1402.
55. Enk AH, Saloga J, Becker D et al. Induction of hapten-specific tolerance by interleukin 10 in vivo. J Exp Med 1994; 179:1397-1402.
56. Ullrich SE. Mechanisms involved in the systemic suppression of antigen-presenting cell function by UV irradiation. Keratinocyte-derived IL-10 modulates antigen-presenting cell function of splenic adherent cells. J Immunol 1994; 152:3410-3416.
57. Schwarz A, Grabbe S, Riemann H et al. In vivo effects of interleukin-10 on contact hypersensitivity and delayed type hypersensitivity reactions. J Invest Dermatol 1994; 103:211-216.
58. Shreedhar V, Giese T, Sung VW et al. A cytokine cascade including prostaglandin E_2, IL-4, and IL-10 is responsible for UV-induced systemic immune suppression. J Immunol 1998; 160:3783-3789.
59. Teunissen MB, Piskin G, Nuzzo S et al. Ultraviolet B radiation induces a transient appearance of IL-4+ neutrophils, which support the development of Th2 responses. J Immunol 2002; 168:3732-3739.
60. Niizeki H, Streilein JW. Hapten-specific tolerance induced by acute, low-dose ultraviolet B radiation of skin is mediated via interleukin-10. J Invest Dermatol 1997; 109:25-30.
61. Mohammad T, Morrison H, HoengEsch H. Urocanic acid photochemistry and photobiology. Photochem Photobiol 1999; 69:115-135.
62. DeFabo EC, Noonan FP. Mechanism of immune suppression by ultraviolet irradiation in vivo. Ist ed. Evidence for the existence of a unique photoreceptor in skin and its role in photoimmunology. J Exp Med 1983; 158:84-89.

63. Shimizu T, Streilein JW. Evidence that ultraviolet B radiation induces tolerance and impairs induction of contact hypersensitivity by different mechanisms. Immunology 1994; 82:140-148.
64. Moodycliffe AM, Norval M, Kimber I et al. Characterization of a monoconal antibody to cis-urocanic acid: Detection of cis-urocanic acid in the serum of irradiated mice by immunoassay. Immunology 1993; 79:265-275.
65. Khalil Z, Townley SL, Grimbaldeston MA et al. cis-Urocanic acid stimulates neuropeptide release from peripheral sensory nerves. J Invest Dermatol 2001; 117:886-891.
66. Misery L. The neuro-immuno-cutaneous system and ultraviolet radiation. Photodermatol Photoimmunol Photomed 2000; 16:78-81.
67. Grabbe S, Bhardwaj RS, Mahnke K et al. alpha-Melanocyte-stimulating hormone induces hapten-specific tolerance in mice. J Immunol 1996; 156:473-478.
68. Hosoi J, Murphy GF, Egan CL et al. Regulation of Langerhans cell function by nerves containing calcitonin gene-related peptide. Nature 1993; 363:159-163.
69. Benrath J, Eschenfelder C, Zimmerman M et al. Calcitonin gene-related peptide, substance P and nitric oxide are involved in cutaneous inflammation following ultraviolet irradiation. Eur J Pharmacol 1995; 293:87-96.
70. Niizeki H, Alard P, Streilein JW. Calcitonin gene-related peptide is necessary for ultraviolet B-impaired induction of contact hypersensitivity. J Immunol 1997; 159:5183-5186.
71. Gillardon F, Moll I, Michel S et al. Calcitonin gene-related peptide and nitric oxide are involved in ultraviolet radiation-induced immunosuppression. Eur J Pharmacol 1995; 293:395-400.
72. Kitazawa T, Streilein JW. Hapten-specific tolerance promoted by calcitonin gene-related peptide. J Invest Dermatol 2000; 115:942-948.
73. Hart PH, Grimbaldeston MA, Swift GJ et al. Dermal mast cells determine susceptibility to ultraviolet B-induced systemic suppression of contact hypersensitivity responses in mice. J Exp Med 1998; 187:2045-2053.
74. Hart PH, Grimbaldeston MA, Swift GJ et al. A critical role for dermal mast cells in cis-urocanic acid-induced systemic suppression of contact hypersensitivity responses in mice. Photochem Photobiol 1999; 70:807-812.
75. Alard P, Niizeki H, Hanninen L et al. Local ultraviolet B irradiation impairs contact hypersensitivity induction by triggering release of tumor necrosis factor-alpha from mast cells. Involvement of mast cells and Langerhans cells in susceptibility to ultraviolet B. J Invest Dermatol 1999; 113:983-990.
76. Alard P, Kurimoto I, Niizeki H et al. Hapten-specific tolerance induced by acute, low-dose ultraviolet B radiation of skin requires mast cell degranulation. Eur J Immunol 2001; 31:1736-1746.
77. Stingl G, Shevach EM. Langerhans cells as antigen-presenting cells. In: Schuler G, ed. Epidermal Langerhans Cells. Boca Raton, Ann Arbor, Boston: CRC Press, 1991; 159-190.
78. Schuler G, Steinman RM. Murine epidermal Langerhans cells mature into potent immunostimulatory dendritic cells in vitro. J Exp Med 1985; 161:526-546.
79. Vermeer M, Streilein JW. Ultraviolet B light-induced alterations in epidermal Langerhans cells are mediated in part by tumor necrosis factor-alpha. Photodermatol Photoimmunol Photomed 1990; 7:258-265.
80. Shimizu T, Streilein JW. Local and systemic consequences of acute, low-dose ultraviolet B radiation are mediated by different immune regulatory mechanisms. Eur J Immunol 1994; 24:1765-1770.
81. Hammerberg C, Duraiswamy N, Cooper KD. Active induction of unresponsiveness (tolerance) to DNFB by in vivo ultraviolet-exposed epidermal cells is dependent upon infiltrating class II MHC^+ $CD11b^{bright}$ monocytic/macrophagic cells. J Immunol 1994; 153:4915-4924.
82. Hammerberg C, Duraiswamy N, Cooper KD. Reversal of immunosuppression inducible through ultraviolet-exposed skin by in vivo anti-CD11b treatment. J Immunol 1996; 157:5254-5261.
83. Hammerberg C, Katiyar SK, Carroll M et al. Activated complement component 3 (C3) is required for ultraviolet induction of immunosuppression and antigenic tolerance. J Exp Med 1998; 187:1133-1138.
84. Trinchieri G. Interleukin-12 and its role in the generation of Th1 cells. Immunol Today 1993; 14:335-338.

85. Riemann H, Schwarz A, Grabbe S et al. Neutralization of interleukin 12 in vivo prevents induction of contact hypersensitivity and induces hapten specific tolerance. J Immunol 1996; 156:1799-1803.
86. Müller G, Saloga J, Germann T et al. IL-12 as mediator and adjuvant for the induction of contact sensitivity in vivo. J Immunol 1995; 155:4661-4668.
87. D'Andrea A, Aste-Amezaga M, Valiante NM et al. Interleukin 10 inhibits human lymphocyte interferon gamma production by suppressing natural killer cell stimulatory factor/IL-12 synthesis in accessory cells. J Exp Med 1993; 178:1041-1048.
88. Schmitt DA, Owen-Schaub L, Ullrich SE. Effect of IL-12 on immune suppression and suppressor cell induction by ultraviolet radiation. J Immunol 1995; 154:5114-5120.
89. Schwarz A, Grabbe S, Aragane Y et al. Interleukin-12 prevents UVB-induced local immunosuppression and overcomes UVB-induced tolerance. J Invest Dermatol 1996; 106:1187-1191.
90. Beissert S, Rühlemann D, Mohammad T et al. IL-12 prevents the inhibitory effects of cis-urocanic acid on tumor antigen presentation by Langerhans cells: Implications for photocarcinogenesis. J Immunol 2001; 167:6232-6238.
91. Ando O, Suemoto Y, Kurimoto M et al. Deficient Th1-type immune responses via impaired CD28 signaling in ultraviolet B-induced systemic immunosuppression and the restorative effect of IL-12. J Dermatol Sci 2000; 24:190-202.
92. Schwarz A, Grabbe S, Grosse-Heitmeyer K et al. Ultraviolet light induced induced immune tolerance is mediated via the CD95/CD95-ligand system. J Immunol 1998; 160:4262-4270.
93. Hill LL, Shreedhar VK, Kripke ML et al. A critical role for Fas ligand in the active suppression of systemic immune responses by ultraviolet radiation. J Exp Med 1999; 189:1285-1293.
94. Applegate LA, Ley RD, Alcalay J et al. Identification of the molecular target for the suppression of contact hypersensitivity by ultraviolet radiation. J Exp Med 1989; 170:1117-1131.
95. Yarosh D, Bucana C, Cox P et al. Localization of liposomes containing a DNA repair enzyme in murine skin. J Invest Dermatol 1994; 103:461-468.
96. Kripke ML, Cox PA, Alas LG et al. Pyrimidine dimers in DNA initiate systemic immunosuppression in UV-irradiated mice. Proc Natl Acad Sci USA 1992; 89:7516-7520.
97. Nishigori C, Yarosh DB, Ullrich SE et al. Evidence that DNA damage triggers interleukin 10 cytokine production in UV-irradiated murine keratinocytes. Proc Natl Acad Sci USA 1996; 93:10354-10359.
98. Kibitel JT, Yee V, Yarosh DB. Enhancement of ultraviolet-DNA repair in denV gene transfectants and T4 endonuclease V-liposome recipients. Photochem Photobiol 1991; 54:753-760.
99. Boonstra A, van Oudenaren A, Baert M et al. Differential ultraviolet-B-induced immunomodulation in XPA, XPC, and CSB DNA repair-deficient mice. J Invest Dermatol 2001; 117:141-146.
100. Stege H, Roza L, Vink AA et al. Enzyme plus light therapy to repair DNA damage in ultraviolet-B-irradiated human skin. Proc Natl Acad Sci USA 2000; 97:1790-1795.
101. Schwarz A, Stander S, Berneburg M et al. Interleukin-12 suppresses ultraviolet radiation-induced apoptosis by inducing DNA repair. Nat Cell Biol 2002; 4:26-31.
102. Schmitt DA, Walterscheid JP, Ullrich SE. Reversal of ultraviolet radiat ion-induced immune suppression by recombinant interleukin-12: Suppression of cytokine production. Immunology 2000; 101:90-96.
103. Aragane Y, Kulms D, Luger TA et al. Down-regulation of interferon gamma-activated STAT1 by UV light. Proc Natl Acad Sci USA 1997; 94:11490-11495.
104. Kulms D, Schwarz T. Ultraviolet radiation inhibits interleukin-2-induced tyrosine phosphorylation and the activation of STAT5 in T lymphocytes. J Biol Chem 2001; 276:12849-12855.
105. Bestak R, Halliday GM. Chronic low-dose UVA irradiation induces local suppression of contact hypersensitivity, Langerhans cell depletion and suppressor cell activation in C3H/HeJ mice. Photochem Photobiol 1996; 64:969-974.
106. Dumay O, Karam A, Vian L et al. Ultraviolet AI exposure of human skin results in Langerhans cell depletion and reduction of epidermal antigen-presenting cell function: Partial protection by a broad-spectrum sunscreen. Br J Dermatol 2001; 144:1161-1168.
107. Garssen J, de Gruijl F, Mol D et al. UVA exposure affects UVB and cis-urocanic acid-induced systemic suppression of immune responses in Listeria monocytogenes-infected Balb/c mice. Photochem Photobiol 2001; 73:432-438.

CHAPTER 12

Dendritic Cells As a Target for Therapeutic Intervention of Contact Hypersensitivity

Akira Takashima, Hiroyuki Matsue, Tadashi Kumamoto, Norikatsu Mizumoto, Akimichi Morita and Mark E. Mummert

Summary

Langerhans cells (LC), a skin-specific member of the dendritic cell (DC) family of antigen presenting cells, play crucial roles in the induction of allergic contact hypersensitivity (CHS) responses. Skin exposure to reactive haptens causes LC emigration from the epidermis as well as their maturation. Once the hapten-loaded, mature LC reach to draining lymph node, they deliver activation signals to hapten-reactive T cells, thereby triggering their clonal expansion. Our working hypothesis is, therefore, that LC serve as a potential target for therapeutic intervention of allergic contact dermatitis (ACD). Indeed, we have recently developed three strategies that are designed to prevent the onset of allergic CHS responses by altering different aspects of LC function. First, we have isolated a 12-mer peptide, termed "Pep-1", which selectively binds to and inhibits the function of hyaluronan. Local administration of Pep-1 in mice prevented hapten-triggered LC migration from the epidermis, thereby inhibiting CHS at the elicitation phase. Secondly, we have converted DC into tolerogenic DC by introducing cDNA encoding CD95L. Upon in vivo administration, the resulting "killer" DC suppressed CHS responses in a hapten-specific manner by eliminating hapten-reactive T cells. Finally, we have observed that LC in CD39-deficient mice expressed no detectable ecto-ATPase/ADPase activities and that these mice exhibited severely impaired CHS. Thus, CD39, which is responsible for LC-associated ecto-ATPase/ADPase activities, may serve as a therapeutic target for allergic contact dermatitis. Taken together, our findings support the concept that allergic CHS can be manipulated experimentally by altering the function of LC.

Introduction

Allergic contact hypersensitivity (CHS) responses are manifested clinically as allergic contact dermatitis (ACD), a common skin disease caused by skin exposure to environmental or industrial allergens. CHS is defined immunologically as T cell-mediated inflammatory responses to reactive haptens. CHS responses differ from conventional delayed-type hypersensitivity responses to foreign protein antigens in that $CD8^+$ T cells act as primary effector leukocytes.[1-3] Langerhans cells (LC) and perhaps dermal dendritic cells (DC) have been considered to act as primary antigen presenting cells for the induction of CHS, at least in the sensitization phase.[4]

Upon skin exposure to reactive haptens, LC migrate to draining lymph nodes, where antigen presentation takes place, and undergo maturational changes including elevated surface

Immune Mechanisms in Allergic Contact Dermatitis, edited by Andrea Cavani and Giampiero Girolomoni. ©2005 Eurekah.com.

expression of MHC class II molecules and costimulatory molecules and production of various cytokines and chemokines. In vivo administration of LC after ex vivo hapten loading leads to successful sensitization as measured by subsequent ear swelling responses to the same hapten, indicating the intrinsic ability of LC to initiate CHS.[5,6] Thus, our central hypothesis is that allergic CHS must be preventable by blocking LC migration, LC maturation, and/or LC-induced T cell activation. In this chapter, we will review our progress in development of new protocols for experimental manipulation of LC function.

Blocking of LC Migration with Pep-1—A Synthetic Peptide Inhibitor of Hyaluronan

Cellular trafficking is controlled at least in part by molecular interaction of adhesion molecules with their corresponding ligands. With respect to LC homing mechanisms, homophilic interaction of E-cadherin is thought to mediate LC retention within multilayered keratinocytes in the epidermal compartment,[7] and α6 integrins appear to promote LC migration through the dermo-epidermal junction.[8] Several lines of evidence now suggest the involvement of CD44 in LC migration: (a) CD44 expression by LC is upregulated by inflammatory stimuli in vitro;[9] (b) LC begin to express CD44v4, v5, v6, and v9 isoforms in vivo following hapten painting;[10] and (c) LC emigration from the epidermis can be blocked by anti-CD44v6 monoclonal antibody (mAb) in the skin organ culture system.[10] Because hyaluronan (HA), a large glycosaminoglycan consisting of repeating disaccharide units of *N*-acetyl glucosamine and glucuronic acid, is expressed in exceptionally large quantities in skin and is known to serve as a primary ligand of CD44,[11,12] we reasoned that LC might employ HA as an adhesive substrate for their CD44-dependent migration. To test this concept, we isolated a peptide inhibitor of HA by using the phage display technique.[13]

Briefly, the M13 phage library expressing random 12-mer peptides fused to the pIII minor coat protein (with the complexity of about 10^9) was incubated on polystyrene plates that had been coated with HA. After removal of unbound phage, hyaluronidase was added to the panning plates to elute only those phage clones that had bound to HA, but not to polystyrene surfaces. After four cycles of panning over HA-coated plates, we isolated and sequenced 19 independent phage clones. Strikingly, despite a theoretical complexity of 10^9 in the starting population, an overwhelming majority (13/19) of the isolated clones expressed an identical peptide motif of GAHWQFNALTVR, which was designated as "Pep-1". This peptide sequence showed no significant homology to any known HA-binding proteins, including CD44.

We next synthesized Pep-1 to study its function. Not only did Pep-1 show significant binding to HA in soluble, immobilized, and cell membrane-associated forms, it also inhibited CD44-dependent adhesion of our XS106 LC-like cell line to HA-coated plates. By contrast, a control peptide designated as random peptide (RP) did not bind to or inhibit the function of HA. A key question then concerned whether one would be able to prevent LC migration using Pep-1. To test this, we injected Pep-1 subcutaneously before topical application of dinitrofluorobenzene (DNFB) and counted the number of LC that remained in the epidermis. A single local administration of Pep-1, but not RP, was found to block this hapten-triggered LC migration almost completely, supporting our hypothesis that LC utilize HA as an adhesive substrate for their migration.

Migration of hapten-pulsed LC to the draining lymph node (where antigen presentation takes place) is the first event for successful sensitization to a topically applied hapten. Thus, we hypothesized further that Pep-1 treatment could be used to prevent allergic CHS responses at the sensitization phase. In fact, mice that were sensitized with DNFB through Pep-1-injected skin sites showed significantly reduced ear swelling responses to DNFB upon challenge. Somewhat unexpectedly, Pep-1 also inhibited ear swelling responses even when administered after DNFB sensitization. Histological examination revealed that Pep-1 reduced the extent of edema and the degree of leukocyte infiltration. Based on these results, we have postulated that HA plays an essential role in two-way trafficking of leukocytes in CHS, i.e., LC emigration from the sensitization sites and skin-directed homing of inflammatory leukocytes to the elicitation sites.[13] Our findings may provide the basis for the future application of Pep-1 and its derivatives for the prevention and treatment of allergic contact dermatitis.

Preclinical efficacies have been well documented for CD44 inhibitors (e.g., anti-CD44 mAb and CD44-Fc fusion proteins) for many inflammatory disease models, such as ACD,[10,14] collagen-induced arthritis,[15] autoimmune type I diabetes,[16] experimental autoimmune encephalomyelitis,[17] and allogeneic skin graft rejection.[18] However, Pep-1 differs fundamentally from the CD44 inhibitors in that it is formulated to block the function of HA, but not of an HA receptor (i.e., CD44). This is of particular importance because CD44 binds not only HA, but also collagens,[19,20] fibronectin,[21] chondroitin sulfates,[22] heparin,[22] heparin sulfate,[22] and serglycins.[23] Conversely, HA binds not only to CD44, but also two additional surface receptors known as the receptor for HA-mediated motility (RHAMM)[24] and LYVE-1.[25] Thus, we believe that Pep-1 may represent a unique pharmacological agent that blocks HA interaction with its diverse receptors.

Elimination of Hapten-Reactive T Cells By CD95L-Transduced "Killer" DC

During antigen presentation, molecular interaction between the MHC/antigenic peptide complex (on DC) and the T cell receptor/CD3 complex (on T cells) ensures antigen specificity, while interaction between costimulatory ligands and their corresponding receptors facilitates full T cell activation. We reasoned that DC that are genetically engineered to over-express a death ligand may deliver apoptotic signals, instead of activation signals, to T cells during antigen presentation. CD95 is known to be expressed by many cell types, primarily after stimulation. For example, resting T cells express no detectable CD95, whereas they acquire its expression rapidly after activation by antigen-pulsed DC, thereby becoming susceptible to CD95/CD95L induced apoptosis.[26,27] We therefore sought to introduce CD95L cDNA into DC to creat "killer" DC.[28]

For this purpose, we employed a stable DC line XS106 which we developed several years ago from the epidermis of newborn A/J mice.[29] Unlike other DC lines of the XS series expressing immature features,[30-34] the XS106 line exhibits mature features, including surface expression of relatively large amounts of IA molecules, CD80, and CD86 and a potent ability to initiate cellular immune responses upon in vivo administration. Using a particle-mediated gene delivery system, we introduced a full length cDNA of mouse CD95L inserted into a pMKITNeo vector (containing the neomycin resistance gene) to XS106 DC. When tested 2 days after transfection, only a small fraction (< 1%) of cells expressed detectable CD95L protein. Thus, we selected stable transfectant clones by the standard limiting dilution microculture in the presence of G418. One of the resulting clones, termed XS106-CD95L, was found to clearly express CD95L mRNA and protein, while maintaining all other surface phenotype of the parental XS106 line. We also selected a control DC clone (termed XS106-neo) after transfection with a pMKITNeo vector alone.

XS106-CD95L clone, but not XS106-neo clone, triggered apoptosis of Jurkat target cells expressing CD95, indicating that CD95L expressed on our killer DC is functionally active. Moreover, XS106-CD95L clone induced apoptosis of an ovalbumin (OVA)-reactive $CD4^+$ T cell clone MH1 generated from syngeneic A/J mice. Most importantly, OVA-pulsed killer DC showed a dramatically augmented ability to kill MH1 T cell targets than did nonpulsed killer DC, with the implication that antigen-specific interaction between DC and T cells is required for optimal delivery of apoptotic signals. Moreover, the addition of anti-CD95L mAb to the apoptosis assay abrogated the cytotoxic potential of killer DC completely.

A/J mice that had received subcutaneous (s.c.) injection of OVA-pulsed XS106-neo DC exhibited significant footpad swelling responses upon subsequent challenge with OVA, whereas mice receiving OVA-pulsed XS106-CD95L DC showed no swelling responses. Furthermore, s.c. administration of OVA-pulsed killer DC, but not control DC, before sensitization with OVA plus complete Freund's adjuvant almost completely inhibited footpad swelling responses to OVA upon challenge. With respect to antigen-specificity, OVA-pulsed killer DC suppressed footpad swelling responses to OVA, but not to an irrelevant antigen, hen egg lysosome (HEL). Conversely, HEL-pulsed killer DC inhibited HEL responses, without affecting OVA responses. These results document the in vivo ability of killer DC to induce antigen-specific immunosuppression.

A single s.c. injection of the parental XS106 DC line or XS106-neo DC clone after in vitro pulsing with DNFB led to successful sensitization in A/J mice, whereas DNFB-pulsed killer DC failed to sensitize the animals. More importantly, administration of DNFB-pulsed killer DC before sensitization inhibited the expression of CHS to DNFB. Once again, killer DC showed no inhibition when tested in the absence of DNFB pulsing. To test the therapeutic efficacy, we next administered DNFB-pulsed killer DC to A/J mice after topical sensitization to DNFB. Significant suppression was observed even when killer DC were given in already sensitized animals. It should be noted that killer DC treatment did not induce long-term tolerance; the above mice that received DNFB-pulsed killer DC after the first sensitization, when tested after the second sensitization, exhibited significant ear swelling responses to DNFB. In summary, hapten-pulsed killer DC inhibit primary CHS responses without causing hapten-specific immunological tolerance.[28]

Our killer DC technology may represent a novel immunosuppressive protocol for selective elimination of pathogenic T cells that recognize a given antigen (i.e., allergen, auto-antigen, allo-antigen). If correct, it may become widely applicable to the treatment of diverse inflammatory disorders in which T cells play causative roles. Several questions, however, remain to be addressed. What is the extent to which antigen-reactive T cells can be eliminated physically by killer DC? Can we use killer DC for the prevention of allo-immune responses? How can we translate such a technology into a clinically applicable form? What is the relative safety? Studies are in progress in our laboratory to address each of these questions (see below).

Working with DO11.10 transgenic mice in which the majority of $CD4^+$ T cells express OVA-reactive, transgenic T cell receptors, we have observed that only partial (up to 40%) elimination is achievable by killer DC technology in the above $CD4^+$ T cell pool.[35] These observations may explain an underlying mechanism for the failure of killer DC to induce long-term tolerance. In an attempt to suppress allogenic immune responses, we have created killer DC-DC "hybrids" by fusing the XS106-CD95L DC clone (derived from A/J mice) with splenic DC freshly isolated from BALB/c mice. The resulting hybrid clones expressed not only CD95L, but also MHC class I and class II molecules of both A/J and BALB/c origins. Repeated intravenous administration of killer DC-DC hybrids significantly delayed the onset of acute graft-versus-host disease in (A/J x BALB/c)F1 mice receiving A/J-derived hematopoietic cells.[36] Although the same killer DC-DC hybrids failed to prevent allogeneic skin graft rejection between the two parental strains (A/J and BALB/c mice), our results support the concept that allo-specific immune responses can be suppressed by genetic and cellular engineering of DC. With regard to safety, systemic administration of agonistic anti-CD95 mAb is known to kill mice rapidly primarily by causing massive apoptosis of hepatocytes.[37] Thus, we considered liver toxicity as a potential risk associated with killer DC treatment. None of the >500 animals receiving killer DC or killer DC-DC treatments showed any apparent adverse effects, except for slightly elevated levels of alanine aminotransferase and γ-glutamyl transpeptidase. Thus, the relative risk for acute liver cytotoxicity appears to be relatively low.[28] Obviously, the killer DC technology in its current form, which requires ex vivo transfection of DC, is not readily applicable to human patients. In this regard, we have been able to target the expression of a foreign gene to epidermal LC by gene gun-mediated delivery of a plasmid DNA under a DC-specific promoter.[38] Moreover, we have been able to entrap migratory LC in a given anatomical location by creating an artificial gradient of LC-attracting chemokine, macrophage inflammatory protein-3β.[39] We anticipate that these technologies of LC-targeted gene expression and LC entrapment may lead to the development of novel in situ protocols for converting epidermal LC into CD95L-transduced killer LC.

LC-Associated Ecto-ATPase/ADPase As a Potential Therapeutic Target

Histo-enzymatic staining of surface ATPase and ADPase activities has been used for >30 years as a useful method to identify LC in skin specimens.[40,41] On the other hand, the molecular identity or function of such ecto-nucleoside triphosphate diphosphohydrolase (NTPDase) activities expressed by LC remains largely unknown. CD39 was originally identified as an obscure activation marker for B cells, T cells, DC, and endothelial cells.[42,43] More recently, CD39 was identified to be responsible for ecto-NTPDase activities expressed on endothelial cells.[44,45] Many ecto-NTPDases have since been isolated and cloned from different cell types and tissues, forming a rapidly growing family of ecto-NTPDases.[46] Endothelial cells release ATP and ADP in response to mechanical shear forces, stretch, changes in osmolarity, oxidative stress, and lipopolysaccharide. Extracellularly release ATP and ADP, in turn, activate ligand-gated ion channel P2X receptors and G protein-coupled P2Y receptors, thereby inducing apoptotic, pro-inflammatory, and thrombotic changes.[47] The P2 receptor signaling pathway is negatively regulated by CD39-dependent hydrolysis of ATP and ADP into AMP.

The most prominent phenotype of CD39-deficient mice is platelet dysfunction characterized by prolonged bleeding times and impaired platelet aggregation.[48] With respect to mechanisms, CD39 deficiency appears to cause $P2Y_1$ receptor desensitization on platelets due to prolonged exposure to its ligand (i.e., ADP). $CD39^{-/-}$ mice show no apparent hematopoietic abnormalities, providing us with a unique opportunity to determine whether CD39 is responsible for LC-associated ecto-NTPDase activities. We have reported recently that $CD39^{-/-}$ mice are indistinguishable from wild-type control mice in the number, distribution, or morphology of epidermal LC defined by their expression of IA molecules and DEC-205. However, neither ecto-ATPase nor ADPase activity was detected in the LC in $CD39^{-/-}$ mice, indicating that CD39 alone accounts for LC-associated ecto-NTPDase activities.[49]

Compared with wild-type mice, $CD39^{-/-}$ mice showed markedly exacerbated skin inflammatory responses to topically applied skin irritant chemicals, croton oil, benzalkonium chloride, and ethyl phenylpropiolate. As to underlying mechanisms, we have observed that keratinocytes release ATP and ADP rapidly upon in vitro exposure to each of the above irritant chemicals, suggesting that LC-associated CD39 plays a protective role against irritant CHS responses by hydrolyzing otherwise proinflammatory ATP and ADP that are released from neighboring keratinocytes.

Allergic CHS to reactive haptens to a reactive hapten oxazolone (OX) were severely attenuated in $CD39^{-/-}$ mice, despite the facts that LC in $CD39^{-/-}$ mice did migrate to lymph nodes and exhibit mature phenotypic features upon hapten painting. On the other hand, OX-loaded, migratory LC isolated from the draining lymph nodes of OX-painted $CD39^{-/-}$ mice showed a significantly diminished ability to activate OX-reactive T cells in vitro as compared to those from the wild-type counterpart. Moreover, bone marrow-derived DC generated from $CD39^{-/-}$ mice, when administered into wild-type mice after hapten pulsing, failed to initiate full allergic CHS responses. These results suggest that CD39 expression by DC is required for primary stimulation of hapten-reactive T cells.

T cells are known to elevate pericellular concentrations of ATP upon activation,[50] while DC have been reported to express several different P2 receptors (e.g., $P2X_1$, $P2X_4$, $P2X_5$, $P2X_7$, $P2Y_1$, $P2Y_2$, $P2Y_4$, $P2Y_5$, $P2Y_6$, $P2Y_{10}$, and $P2Y_{11}$) and to respond to nucleotide stimulation by pore formation, Ca^{2+} influx, phenotypic maturation, chemotactic migration, apoptosis, and cytokine production.[51-57] Thus, it is conceivable that molecular interaction between ATP (produced by T cells) and P2 receptors (expressed on DC) is functionally involved in antigen presentation. Our current hypothesis is, therefore, that CD39 on DC regulates the ATP-mediated DC-T cell communication by preventing P2 receptor desensitization, as has been documented for $CD39^{-/-}$ platelets.[48] In fact, bone marrow-derived DC generated from $CD39^{-/-}$ mice failed to respond to ATP stimulation and this unresponsiveness was reversed by pretreatment of $CD39^{-/-}$ DC with soluble NTPDase. Although molecular identities of the P2 receptors that are functionally paralyzed in $CD39^{-/-}$ DC remain to be determined, our observations suggest a previously unrecognized role of DC-associated CD39 in regulating nucleotide-mediated intercellular communication between DC and T cells during antigen presentation.

In summary, we have identified CD39 to be responsible for LC-associated ecto-NTPDase activities and demonstrated opposing outcomes of CD39 deficiency in skin inflammation (e.g., irritant CHS responses) and immune responsiveness (e.g., allergic CHS responses). These observations may form conceptual frameworks for the development of a new class of therapeutic agents that are designed to alter the function of CD39 or relevant P2 receptors on DC.[49]

References

1. Gocinski BL, Tigelaar RE. Roles of $CD4^+$ and $CD8^+$ T cells in murine contact sensitivity revealed by in vivo monoclonal antibody depletion. J Immunol 1990; 144:4121-4128.
2. Bour H, Peyron E, Gaucherand M et al. Major histocompatibility complex class I-restricted CD8+ T cells and class II restricted CD4+ T cells, respectively, mediate and regulate contact sensitivity to dinitrofluorobenzene. Eur J Immunol 1995; 25(11):3006-3010.
3. Bouloc A, Cavani A, Katz SI. Contact hypersensitivity in MHC class II-deficient mice depends on CD8 T lymphocytes primed by immunostimulating Langerhans cells. J Invest Dermatol 1998; 111:44-49.
4. Grabbe S, Schwarz T. Immunoregulatory mechanisms involved in elicitation of allergic contact hypersensitivity. Immunol Today 1998; 19:37-44.
5. Tamaki K, Fujiwara H, Katz SI. The role of epidermal cells in the induction and suppression of contact sensitivity. J Invest Dermatol 1981; 76:275-278.
6. Sullivan S, Bergstresser PR, Tigelaar RE et al. Induction and regulation of contact hypersensitivity by resident, bone-marrow derived, dendritic epidermal cells: Langerhans cells and Thy-1+ epidermal cells. J Immunol 1986; 137:2460-2467.
7. Tang A, Amagai M, Granger LG et al. Adhesion of epidermal Langerhans cells to keratinocytes mediated by E-cadherin. Nature 1993; 361:82-85.
8. Price AA, Cumberbatch M, Kimber I et al. α_6 integrins are required for Langerhans cell migration from the epidermis. J Exp Med 1997; 186:1725-1735.
9. Osada A, Nakashima H, Furue M et al. Up regulations of CD44 expression by tumor necrosis factor-α is neutralized by interleukin 10 in Langerhans cells. J Invest Dermatol 1995; 105:124-127.
10. Weiss JM, Sleeman J, Renkl AC et al. An essential role for CD44 variant isoforms in epidermal Langerhans cell and blood dendritic cell function. J Cell Biol 1997; 137:1137-1147.
11. Aruffo A, Stamenkovic I, Melnick M et al. CD44 is the principal cell surface receptor for hyaluronate. Cell 1990; 61:1303-1313.
12. Miyake K, Underhill CB, Lesley J et al. Hyaluronate can function as a cell adhesion molecule and CD44 participates in hyaluronate recognition. J Exp Med 1990; 172:69-75.
13. Mummert ME, Mohamadzadeh M, Mummert DI et al. Development of a peptide inhibitor of hyaluronan-mediated leukocyte trafficking. J Exp Med 2000; 192:769-779.
14. Camp RL, Scheynius A, Johansson C et al. CD44 is necessary for optimal contact allergic responses but is not equired for normal leukocyte extravasation. J Exp Med 1993; 178:497-507.

15. Mikecz K, Brennan FR, Kim JH et al. Anti-CD44 treatment abrogates tissue oedema and leukocyte infiltration in murine arthritis. Nature Med 1995; 1(6):558-563.
16. Weiss L, Slavin S, Reich S et al. Induction of resistance to diabetes in nonobese diabetic mice by targeting CD44 with specific monoclonal antibody. Proc Natl Acad Sci USA 2000; 97:285-290.
17. Brocke S, Piercy C, Steinman L et al. Antibodies to CD44 and integrin α4, but not L-selectin, prevent central nervous system inflammation and experimental encephalomyelitis by blocking secondary leukocyte recruitment. Proc Natl Acad Sci USA 1999; 96(12):6896-6901.
18. Seiter S, Weber B, Tilgen W et al. Down-modulation of host reactivity by anti-CD44 in skin transplantation. Transplantation 1998; 66(6):778-791.
19. Ehnis T, Dieterich W, Bauer M et al. A chondroitin/dermatan sulfate form of CD44 is a receptor for collagen XIV (undulin). Exp Cell Res 1996; 229(2):388-397.
20. Knutson JR, Iida J, Fields GB et al. CD44/chondroitin sulfate proteoglycan and α2β1 integrin mediate human melanoma cell migration on type IV collagen and invasion of basement membranes. Mol Biol Cell 1996; 7:383-396.
21. Jalkanen S, Jalkanen M. Lymphocyte CD44 binds the COOH-terminal heparin-binding domain of fibronectin. J Cell Biol 1992; 116(3):817-825.
22. Sleeman JP, Kondo K, Moll J et al. Variant exons v6 and v7 together expand the repertoire of glycosaminoglycans bound by CD44. J Biol Chem 1997; 272:31837-31844.
23. Toyama-Sorimachi N, Kitamura F, Habuchi H et al. Widespread expression of chondroitin sulfate-type serglycins with CD44 binding ability in hematopoietic cells. J Biol Chem 1997; 272(42):26714-26719.
24. Hardwick C, Hoare K, Owens R et al. Molecular cloning of a novel hyaluronan receptor that mediates tumor cell motility. J Cell Biol 1992; 117(6):1343-1350.
25. Banerji S, Ni J, Wang SX et al. LYVE-1, a new homologue of the CD44 glycoprotein, is a lymph-specific receptor for hyaluronan. J Cell Biol 1999; 144(4):789-801.
26. van Parijs LV, Abbas AK. Role of Fas-mediated cell death in the regulation of immune responses. Curr Opin Immunol 1996; 8:355-361.
27. Matiba B, Mariani SM, Krammer PH. The CD95 system and the death of a lymphocyte. Immunology 1997; 9:59-68.
28. Matsue H, Matsue K, Walters M et al. Induction of antigen-specific immunosuppression by CD95L cDNA-transfected "killer" dendritic cells. Nature Med 1999; 5:930-937.
29. Timares L, Takashima A, Johnston SA. Quantitative analysis of the immunopotency of genetically transfected dendritic cells. Proc Natl Acad Sci USA 1998; 95:13147-13152.
30. Xu S, Ariizumi K, Caceres-Dittmar G et al. Successive generation of antigen-presenting, dendritic cell lines from murine epidermis. J Immunol 1995; 154:2697-2705.
31. Xu S, Ariizumi K, Edelbaum D et al. Cytokine-dependent regulation of growth and maturation in murine epidermal dendritic cell lines. Eur J Immunol 1995; 25:1018-1024.
32. Takashima A, Edelbaum D, Kitajima T et al. Colony-stimulating factor-1 secreted by fibroblasts promotes the growth of dendritic cell lines (XS series) derived from murine epidermis. J Immunol 1995; 154:5128-5135.
33. Kitajima T, Caceres-Dittmar G, Tapia FJ et al. T cell-mediated terminal maturation of dendritic cells: Loss of adhesive and phagocytotic capacities. J Immunol 1996; 157:2340-2347.
34. Kitajima T, Ariizumi K, Bergstresser PR et al. T cell-dependent loss of proliferative responsiveness to colony-stimulating factor-1 by a murine epidermal-derived dendritic cell line, XS52. J Immunol 1995; 155:5190-5197.
35. Kusuhara M, Matsue K, Edelbaum D et al. Killing of naive T cells by CD95L-transfected dendritic cells (DC): In vivo study using killer DC-DC hybrids and $CD4^+$ T cells from DO11.10 mice. Eur J Immunol 2002; 32(4):1035-1043.
36. Matsue H, Matsue K, Kusuhara M et al. Immunosuppressive properties of CD95L-transduced "killer" hybrids created by fusing donor- and recipient-derived dendritic cells. Blood 2001; 98:3465-3472.
37. Ogasawara J, Watanabe-Fukunaga R, Adachi M et al. Lethal effect of the anti-Fas antibody in mice. Nature 1993; 364:806-809.
38. Morita A, Ariizumi K, Ritter R III et al. Development of a Langerhans cell-targeted gene therpay format using a dendritic cell-specific promoter. Gene Therapy 2001; 8:1729-1737.

39. Kumamoto T, Huang EK, Paek HJ et al. Induction of tumor-specific protective immunity by in situ Langerhans cell vaccine. Nature Biotechnol 2002; 20:64-69.
40. Wolff K, Winkelmann RK. Ultrastructural localization of nucleoside triphosphatase in Langerhans cells. J Invest Dermatol 1967; 48(1):50-54.
41. Chaker MB, Tharp MD, Bergstresser PR. Rodent epidermal Langerhans cells demonstrate greater histochemical specificity for ADP than for ATP and AMP. J Invest Dermatol 1984; 82(5):496-500.
42. Maliszewski CR, Delespesse GJ, Schoenborn MA et al. The CD39 lymphoid cell activation antigen. Molecular cloning and structural characterization. J Immunol 1994; 153(8):3574-3583.
43. Kansas GS, Wood GS, Tedder TF. Expression, distribution, and biochemistry of human CD39. Role in activation-associated homotypic adhesion of lymphocytes. J Immunol 1991; 146(7):2235-2244.
44. Kaczmarek E, Koziak K, Sevigny J et al. Identification and characterization of CD39/vascular ATP diphosphohydrolase. J Biol Chem 1996; 271(51):33116-33122.
45. Wang TF, Guidotti G. CD39 is an ecto-(Ca2+,Mg2+)-apyrase. J Biol Chem 1996; 271(17):9898-9901.
46. Zimmermann H. Extracellular metabolism of ATP and other nucleotides. Naunyn Schmiedebergs Arch Pharmacol 2000; 362(4-5):299-309.
47. Ralevic V, Burnstock G. Receptors for purines and pyrimidines. Pharmacol Rev 1998; 50(3):413-492.
48. Enjyoji K, Sevigny J, Lin Y et al. Targeted disruption of cd39/ATP diphosphohydrolase results in disordered hemostasis and thromboregulation. Nature Med 1999; 5(9):1010-1017.
49. Mizumoto N, Kumamoto T, Robson SC et al. CD39 is the dominant Langerhans cell-associated ecto-NTPDase: Modulatory roles in inflammation and immune respoonsiveness. Nature Med 2002; 8:358-365.
50. Filippini A, Taffs RE, Sitkovsky MV. Extracellular ATP in T-lymphocyte activation: Possible role in effector functions. Proc Natl Acad Sci USA 1990; 87(21):8267-8271.
51. Mutini C, Falzoni S, Ferrari D et al. Mouse dendritic cells express the $P2X_7$ purinergic receptor: Characterization and possible participation in antigen presentation. J Immunol 1999; 163:1958-1965.
52. Liu QH, Bohlen H, Titzer S et al. Expression and a role of functionally coupled P2Y receptors in human dendritic cells. FEBS Lett 1999; 445(2-3):402-408.
53. Coutinho-Silva R, Persechini PM, Bisaggio RD et al. P2Z/P2X7 receptor-dependent apoptosis of dendritic cells. Am J Physiol 1999; 276(5 Pt 1):C1139-C1147.
54. Berchtold S, Ogilvie AL, Bogdan C et al. Human monocyte derived dendritic cells express functional P2X and P2Y receptors as well as ecto-nucleotidases. FEBS Lett 1999; 458(3):424-428.
55. Marriott I, Inscho EW, Bost KL. Extracellular uridine nucleotides initiate cytokine production by murine dendritic cells. Cell Immunol 1999; 195(2):147-156.
56. Nihei OK, de Carvalho AC, Savino W et al. Pharmacologic properties of $P_{2Z}/P2X_7$ receptor characterized in murine dendritic cells: role on the induction of apoptosis. Blood 2000; 96(3):996-1005.
57. Ferrari D, La Sala A, Chiozzi P et al. The P2 purinergic receptors of human dendritic cells: Identification and coupling to cytokine release. FASEB J 2000; 14(15):2466-2476.

Chapter 13

Evidence-Based Answers to Clinical Questions in Allergic Contact Dermatitis

Whitney A. High and Ponciano D. Cruz

Summary

Four issues of current relevance to clinical contact dermatitis are discussed in the context of basic knowledge regarding immunopathogenesis: What differentiates contact allergens from contact irritants? What governs sensitization versus tolerance? How does patch test reactivity relate to disease? What is the role of patch tests to house dust mite allergen in managing patients with atopic dermatitis?

In 1963, Coombs and Gel described four effector mechanisms of immunological injury, including type IV hypersensitivity as exemplified by allergic contact dermatitis.[1] Four decades hence, our knowledge of how immune responses are generated at cellular and molecular levels has expanded exponentially, yet much remains unknown regarding the pathogenesis of allergic contact dermatitis. Why are some individuals allergic to poison ivy, or to nickel, or to rubber components, whereas others are not? How is it that a patient who has never reacted to a substance to which he/she is exposed constantly, only now presents with a severe skin reaction to that substance? What accounts for divergent responses to the same patch test allergen applied in different clinics or at different time periods?

In this final chapter, we will integrate selected issues of current relevance to clinical contact dermatitis with basic knowledge regarding immunopathogenesis. To achieve this goal, we have posed four fundamental questions that prompt a review of existing scientific evidence, some covered in more detail in previous chapters.

What Differentiates Contact Allergens from Contact Irritants?

Long before we knew about antigen-presenting cells (APC) and cytokines, allergic contact dermatitis (ACD) was well-recognized as a delayed-in-time skin eruption mediated by T cells; the so-called type IV hypersensitivity response.[1] Its designation as a cell-mediated disorder arose from the observation that naive animals could be induced to express the skin disorder if infused with the cellular (but not the serum) component of blood from sensitized animals. ACD is a delayed response because the inflammatory reaction becomes manifest days after exposure to the allergen. Thus, diagnostic patch tests are read at 48 hours and beyond after application, whereas prick or scratch tests (which assay for IgE-mediated responses) are evaluated immediately after the procedure.

Key to understanding the pathogenesis of ACD is appreciation of the discriminative nature in which chemicals can serve as contact antigens (haptens). Experimental data suggests

Immune Mechanisms in Allergic Contact Dermatitis, edited by Andrea Cavani and Giampiero Girolomoni. ©2005 Eurekah.com.

that, at the very least, three sequential criteria are required for a chemical to act as a contact allergen. First, the chemical must be able to penetrate the stratum corneum and reach the living cells of the skin. Second, the chemical must be converted into an antigenic (protein-reactive) form that can bind to the antigen receptor of responder T cells (precursor cells in the case of a primary sensitization and memory cells in the case of subsequent exposures). Third, the bound antigen should be capable of activating responder T cells and produce clonal proliferation.

Two physicochemical properties promote transit of chemicals through the stratum corneum: small molecular size (< 500 daltons) and lipid solubility. Thus, it is no surprise that most haptens are small, lipophilic molecules. However, many small, lipophilic chemicals that can penetrate the stratum corneum do not cause ACD. Chemicals that reach the epidermis (and dermis) may bind to different cell-associated and cell-free proteins to form hapten-protein conjugates that are potentially antigenic. One pathway for these hapten-protein conjugates is to be taken up (endocytosed) by APC and degraded (processed) intracellularly into hapten-peptide conjugates. Subsequent critical events include association of processed peptides with MHC class II molecules and their presentation on the surface of APC. A second pathway is for chemicals to conjugate directly with protein moieties already expressed on the APC surface, either the MHC molecules themselves (class I or class II) or peptides bound to these molecules, thereby circumventing endocytosis and intracellular processing altogether. It is likely that both pathways are used, and that depending on the specific hapten, one or the other pathway predominates.[2-6] Antigens presented in association with MHC class I molecules are recognized by $CD8^+$ T cells, whereas antigens presented in association with MHC class II molecules are recognized by $CD4^+$ T cells. Because contact sensitizers may be presented in the context of either class I or class II molecules, both subpopulations of T cells (CD4 and CD8) may serve mediators of allergic contact dermatitis, as has been shown for nickel.[2-6] To complicate matters, clones of $CD8^+$ cytotoxic T cells from nickel-allergic patients that do not depend on expression of MHC molecules have also been identified.[7]

Langerhans cells (LCs), the principal APC within the epidermis, are thought to be responsible in great part for presentation of contact allergens. Indeed, allergens may be distinguished from irritants based on their effects on LCs after topical application. LCs in allergen-treated skin have more endocytic organelles, up-regulated expression of MHC class II and IL-1β, augmented APC function for syngeneic and allogeneic T-cell proliferation, and increased migration to draining lymph nodes.[8-11]

What Governs Sensitization Versus Tolerance?

Current dogma holds that the host's initial encounter(s) with a specific antigen is crucial in determining whether sensitization or tolerance ensues. Thus, in one scenario, antigen-specific T-cell activation triggers T-cell proliferation leading to cascading events that culminate in a sensitizing (allergic) response. In this scenario, the host develops an inflammatory skin eruption (ACD) that becomes successively worse with each exposure to the same allergen. Alternatively, the first encounter(s) with antigen may produce tolerance or an anergic response. In this scenario, "activation" of the precursor T cells recognizing the antigen results in these cells' death or renders these cells incapable of proliferating even if restimulated by later exposures.

A concept referred to as the 'danger model' may explain whether sensitization or tolerance ensues. For example, certain microbes inhabit the skin surface as commensals, well tolerated by the immune system. However, when these microbes overgrow or invade, pro-inflammatory molecules including cytokines (innate immunity) may be released by skin cells in response to the perceived danger. This inflammatory response may be critical to activating T cells specific for the antigen (acquired immunity).

Observations relevant to ACD support the danger model. Using dinitrochlorobenzene (DNCB) as the allergen, it was discovered that dose per unit area is a critical parameter for sensitization.[12] On the one hand, this could be interpreted as a threshold concentration required for antigenic signaling. On the other hand, supporters of the danger model believe that hapten concentration relates to a need for low-grade irritation below which tolerance is the likely outcome. Consistent with this concept is the fact that most allergens are also irritants.

Cytokines secreted in response to irritation may condition whether sensitization or tolerance ensues. Gleichmann et al demonstrated that nickel in it lowest oxidation state (Ni^{++}) failed to sensitize mice.[13] However, when Ni^{++} was coadministered with a strong irritant, sensitization occurred, presumably aided by inflammation induced by the irritant and by production of reactive-oxygen species responsible for oxidizing nickel to higher states. Likewise, Ni^{++} and hydrogen peroxide applied together led to production of Ni^{+++} and contact sensitization.

Inflammation from sources other than chemical irritation may also foster sensitization. For example, the antibiotic neomycin, is a weak irritant or allergen, but its application on inflamed skin favors sensitization.[14] The same mechanism may explain the propensity for developing ACD to topical steroids when used to treat stasis dermatitis.[15-17] Thus, a pure antigenic signal from a hapten may favor tolerance, while an antigenic signal administered in conjunction with inflammation or other danger signals may produce sensitization.

How Does Patch Test Reactivity Relate to Disease?

Patch testing remains the clinical gold standard for confirming ACD and for identifying offending contact allergens. Yet, this bioassay is fraught with many limitations, which are highlighted in the cases of gold and thimerosal.

Gold is a ubiquitous metal that has been used throughout the ages as currency, personal adornment, in dental apparatus and as an ingredient of medicaments. Metallurgically, gold is relatively inert and difficult to solubilize,[18] leading many to believe that true allergic reactions to gold are rare. However, in the late-1980's, reports of contact dermatitis to gold jewelry began to be reported.[19,20] In addition to dermatitis at points of jewelry contact, eyelid dermatitis was frequently associated with contact allergy to gold.[21-23] Oral lichen planus and lichenoid lesions have also been described in association with gold dental work.[24,25] Systemic allergy to gold can occur with several patients developing lichenoid eruptions following ingestion of gold-containing alcoholic beverages or following therapeutic intramuscular administration.[26,27]

Bjorker et al published the first series of patients routinely tested to gold, noting that 8.6% of 823 patients tested positively.[28] Other studies have shown rates of positive gold reactions ranging from 3 to 13%.[29-36] Thus, sensitization to gold is more common than previously believed.

Although several studies demonstrated a statistically increased rate of hypersensitivity in people with gold dental work,[37-39] none have shown an increase in oral dermatitis in those sensitized to gold, prompting many clinical investigators to interpret these examples as positive, but often irrelevant patch test results.

Thimerosal (also known as merthiolate) is a preservative used in topical drugs, cosmetics, and vaccines. It is composed of two distinct components: an organic ethylmercuric chloride molecule and thiosalicylate, with ACD attributed to either constituent.[40,41] The North American Contact Dermatitis Group (NACDG) reported that thimerosal was the fifth most common allergen, inducing reactions in 11% of 4,087 patients.[42] However, thimerosal was clinically relevant in only 17% of cases, placing it last in relevance among 50 allergens in this group's screening tray. For this reason, many prominent clinicians in the United States have argued to drop thimerosal from standard screening trays, as has already been done by the European Environmental and Contact Dermatitis Research Group (EECDRG).[43]

Thimerosal's presence as a preservative in many commonly administered vaccines complicates the discussion since many epidemiological studies have shown thimerosal sensitivity to be increasing, particularly in younger individuals who have undergone vaccination.[44-46] Recently, Audicana et al used a series of three intramuscular challenges administered every other day in increasing amounts (0.1, 0.5, 1.0 ml) of thimerosal at a concentration similar to that in vaccines (100 ug/ml). They showed that thimerosal was well tolerated in 52 of 57 patients who were patch test reactive to thimerosal.[47] Thus, the current consensus is that vaccines containing thimerosal are acceptable even for individuals with known hypersensitivity to thimerosal.

What is the Role of Patch Tests to House Dust Mite Allergen in Managing Patients with Atopic Dermatitis?

Atopic dermatitis is a common cause of morbidity affecting 7 to 17% of children,[48,49] and to a lesser extent in adults. The role of house dust mite allergens in asthma and allergic rhinitis is well accepted, but its contribution to the pathophysiology of atopic dermatitis is a matter of debate. Exacerbation of atopic dermatitis by products produced by the house dust mites was first proposed by Tuft in 1949.[50] A few authors have also described dramatic clinical improvement of atopic dermatitis following implementation of mite elimination strategies.[50-52] Mitchell et al showed that purified mite allergen produced a hypersensitivity reaction on treated skin of atopic individuals.[53] More recently, Tupker et al demonstrated worsening of atopic dermatitis following exposure to intrabronchial mite allergens.[54]

On the other hand, many of the above mentioned studies were small and/or uncontrolled, and association does not necessarily imply causality. For example, the increased transepidermal water loss and copious shedding of squames (on which the mites feed) that are commonly present in severe atopic dermatitis may support an enhanced mite population in the bedroom/mattress environment of such individuals.[55] Others have cited lack of improvement with mite desensitization therapy, the inability of a standard mite prick test to induce dermatitis, the increased prevalence of atopic dermatitis in dry and northern latitudes where mite populations are sparse, or the difficulty in producing a truly dust-free environment as evidence against such direct involvement.[56,57]

Some investigators have performed epicutaneous testing of atopic patients using a variety of mite-derived antigens; positive results were reported in 21-77% of patients.[58-61] Interestingly, patients with an aeroallergen-type distribution (face, neck, hands, feet) showed, a significantly higher rate of reactivity compared with patients with a nonaeroallergen distribution.[62,63]

Chemotechnique Diagnostics (Malmo, Sweden) produces a standardized dust mite allergen mixtue (D. pteronyssinus and D. farinae, 50:50 mix) in petrolatum in two concentrations (20 and 30%). Concentrations above these levels are considered too irritating.[64] Some have proposed testing of atopic patients with the above mixtures and initiation of avoidance strategies if reactivity is demonstrated.[65] Others have reported a low rate of clinical relevance in larger groups of individuals with dermatitis.[66] Still others have found that the 20 and 30% concentrations elicit many decrescendo reactions characteristic of irritation rather than allergy, and have demonstrated that a 0.1% dilution more appropriately differentiated rates of positivity among healthy controls, atopic dermatitis and respiratory atopy.[67]

The controversy surrounding the role of house dust mites in atopic dermatitis is far from settled. Future investigation focusing on the clinical relevance of epicutaneous testing to more appropriately predict those who would benefit from avoidance—such as a case-control study of "atopic" and "nonatopic" groups—is necessary and will provide valuable clinical information.

References

1. Coombs RRA, Gell PGH. Classification of allergic reaction responsible for clinical hypersensitivity and disease. In: Gell PGH, Coombs RRA, Lachmann PJ, eds. Clinical aspects of immunology. Oxford: Blackwell Scientific Publications, 1975.
2. Kapsenberg ML, Wierenga EA, Bos JD et al. Functional subsets of allergen-reactive human CD4+ T cells. Immunol Today 1991; 2:392-5.
3. Kapsenberg ML, Wierenga EA, Stiekema FEM et al. Th1 lymphokine production profiles of nickel-specific CD4+ T-lymphocyte clones from nickel contact allergic and nonallergic individuals. J Invest Dermatol 1992; 98:59-63.
4. Gocinski BL, Tigelaar RE. Roles of $CD4^+$ and $CD8^+$ T cells in murine contact sensitivity revealed by in vivo monoclonal antibody depletion. J Immunol 1990; 144:4121-8.
5. Cavani A, Mei D, Guerra E. Patients with allergic contact dermatitis to nickle and nonallergic individuals display different nickel-specific T cell responses:Evidence for the presence of effector CD8+ and regulatory CD4+ T cells. J Invest Dermatol 1998; 111:621-8.
6. Cavani A, Albanesi C, Tiaidl C. Effector and regulatory T cells in allergic contact dermatitis. Trends Immnol 2001; 22:118-20.
7. Moulon C, Wild D, Dormoy A et al. MHC-dependent and MHC-independent activation of human nickel-specific $CD8^+$ cytotoxic T-cells from allergic donors. J Invest Dermatol 1998; 111:360-6.
8. Kolde G, Knop J. Different cellular reaction patterns of epidermal Langerhans cells after application of contact sensitizing, toxic and tolerogenic compounds. A comparative ultrastructural and morphometric time-course analysis. J Invest Dermatol 1987; 89:19.
9. Aiba S, Katz SI. Phenotypic and functional characteristics of in vivo-activated Langerhans cells. J Immunol 1990; 145:2791-6.
10. Enk AH, Katz SI. Early molecular events in the induction phase of contact sensitivity. Proc Natl Acad Sci USA 1992; 89:1398.
11. Kripke ML, Munn CG, Jeevan A et al. Evidence that cutaneous antigen-presenting cells migrate to regional lymph nodes during contact sensitization. J Immunol 1990; 145:2833-8.
12. White SJ, Friedmann PS, Moss C. The effect of altering area of application and dose per unit area on sensitization to DNCB. Br J Dermatol 1986; 115:663-8.
13. Artik A, von Vultee C, Gleichmann E. Nickel allergy in mice:Enhanced sensitization capacity of nickel at higher oxidation states. J Immunol 1999; 163:1143-52.
14. Smith HR, McFadden JP, Basketter DA. Contact allergy, irritancy, and "danger". Contact Dermatitis 2000; 42:123-7.
15. Fraki JE, Peltonen L, Hopsu-Havu VK. Allergy to various components of topical preparations in stasis dermatitis and leg ulcer. Contact Dermatitis 1979; 5:97-100.
16. Wilkinson S, English J. Hydrocortisone sensitivity: Clinical features of 59 cases. J Am Acad Dermatol 1992; 27:683-7.
17. Wilkinson SM. Hypersensitivity to topical corticosteroids. Clin Exp Dermatol 1994; 19:1-11.
18. Flint GN. A metallurgical approach to metal contact dermatitis. Contact Dermatitis 1998; 39:213-21.
19. Rapson WS. Skin contact with gold and gold alloys. Contact Dermatitis 1985; 13:56-65.
20. Fisher AA. Allergic dermal contact dermatitis due to gold earrings. CUTIS 1987; 39:473-5.
21. Fowler Jr JF. Allergic contact dermatitis to gold. Arch Dermatol 1988; 124:181-2.
22. Bruze M, Edman B, Bjorkner B. Clinical relevance of contact allergy to gold sodium thiosulfate. J Am Acad Dermatol 1994; 31:579-83.
23. Ehrich A, Belsito D. Allergic contact dermatitis to gold. CUTIS 2000; 65:323-6.
24. Koch P, Bahmer F. Oral lesions and symptoms related to metals used in dental resotrations:A clinical, allergological, and histologic study. J Am Acad Dermatol 1999; 41:422-30.
25. Scalf LA, Fowler Jr JF, Morgan KW. Dental metal allergy in patients with oral, cutaneous, and genital lichenoid reactions. Am J Cont Dermatol 2001; 36:841-4.
26. Russell MA, Langley M, Truett III AP. Lichenoid dermatitis after consumption of gold-containing liquor. J Am Acad Dermatol 1997; 36:841-4.

27. Russell MA, King Jr LE, Boyd AS. Lichen planus after consumption of a gold-containing liquor. N Engl J Med 1996; 334(9):603.
28. Bjorker B, Bruze M, Moller H. High frequency of contact allergy to gold sodium thiosulfate:An indication of gold allergy? Contact Dermatitis 1994; 30:144-51.
29. Fleming C, Forsyth A, MacKie R. Prevalence of gold contact hypersensitivity in the west of Scotland. Contact Dermatitis 1997; 36:302-4.
30. Sabroe RA, Sharp LA, Peachey RD. Contact allergy to gold sodium thiosulfate. Contact Dermatitis 1996; 34:345-8.
31. Koch P, Kiehn M, Frosch PJ. Epicutaneous testing with gold salts:Two multicenter studies of the german contact allergy group. Hautarzt 1997; 48:812-6.
32. McKenna KE, Dolan O, Walsh MY. Contact allergy to gold sodium thiosulfate. Contact Dermatitis 1995; 32:143-6.
33. Silva R, Pereira F, Bordalo O. Contact allergy to gold sodium thiosulfate:A comparative study. Contact Dermatitis 1997; 37:78-81.
34. Leow YH, Ng SK, Goh CL. A preliminary study of gold sensitization in Singapore. Contact Dermatitis 1998; 38:169-70.
35. Trattner A, David M. Gold sensitivity in Israel - consecutive patch test results. Contact Dermatitis 2000; 42:301-2.
36. Lee AY, Eun HC, Kim HO. Multicenter study of the frequency of contact allergy to gold. Contact Dermatitis 2001; 45:214-6.
37. Schaffran RM, Storrs FJ, Schalock P. Prevalence of gold sensitivity in asymptomatic individuals with gold dental restoration. Am J Cont Dermatol 1999; 201-6.
38. Vamnes JS, Morken T, Helland S. Dental gold alloys and contact hypersensitivity. Contact Dermatitis 2000; 42:128-33.
39. Ahlgren C, Ahnlide I, Bjorkner B. Contact allergy to gold is correlated to dental gold. Acta Derm Venereol 2002; 82:41-4.
40. Pirker C, Moslinger T, Wantke F, Gotz M, Jarisch R. Ethylmercuric chloride: the responsible agent in thimerosal hypersensitivity. Contact Dermatitis 1993; 29:152-3.
41. Cirne de Castro JL, Freitas JP, Menezes Brandao F. Sensitivity to thimerosal and photosensitivity to piroxicam. Contact Dermatitis 1991; 24:22-6.
42. Marks JG, Belsito DV, DeLeo VA. North american contact dermatitis group patch-test results, 1996-1998. Arch Dermatol 2000; 136:x.
43. Belsito DV. Thimerosal:Contact (non)allergen of the year. Am J Cont Dermatol 2001; 13:1-2.
44. Forstrom L, Hannuksela M, Kousa M. Timerosal hypersensitivity and vaccination. Cont Dermatitis 1980; 6:241-5.
45. Wantke F, Hemmer W, Jarisch R. Patch test reactions in children, adults, and the elderly. Cont Dermatitis 1996; 34:316-9.
46. Goncalo S, Goncalo M, Azenha A. Allergic contact dermatitis in children. Cont Dermatitis 1992; 26:112-5.
47. Audicana MT, Munoz D, Dolores del Pozo M. Allergic contact dermatitis from mercury antiseptics and derivatives:Study protocol of tolerance to intramuscular injections of thimerosal. Am J Cont Dermatol 2002; 13:3-9.
48. The international study of asthma and allergice in childhood (ISAAC) steering committee. Worldwide prevalence of symptoms of asthma, allergic rhinoconjunctivitis and atopic eczema. Lancet 1998; 351:1225-32.
49. Laughter D, Istvan JA, Tofte SJ et al. The prevalence of atopic dermatitis in oregon schoolchildren. J Am Acad Dermatol 2000; 43(4):649-55.
50. Tuft LA. Importance of inhalant allergens in atopic dermatitis. J Invest Dermatol 1949; 12:211-9.
51. Platts-Mills TA, Mitchell EB, Rowntree S et al. The role of dust mite allergens in atopic dermatitis. Clin Exp Dermatol 1983; 8(3):233-47.
52. Tan BB, Weald D, Strickland I et al. Double-blind controlled trial of effect of housedust-mite allergen avoidance on atopic dermatitis. Lancet 1996; 347(8993):15-8.
53. Mitchell EB, Crow J, Chapman MD et al. Basophils in allergen-induced patch test sites in atopic dermatitis. Lancet 1982; 1(8264):127-30.

54. Tupker RA, de Monchy JG, Coenraads PJ et al. Induction of atopic dermatitis by inhalation of house dust mite. J Allergy Clin Immunol 1996; 97(5):1064-70.
55. Werner Y, Lindberg M. Transepidermal water loss in dry and clinically normal skin in patients with atopic dermatitis. Acta Derm Venereol 1985; 65(2):102-5.
56. Beck HI, Bjerring P. House dust mites and human dander. Allergy 1987; 42(6):471-2.
57. Beltrani V, Hanifin J. Atopic dermatitis, house dust mites, and patch testing. Am J Contact Dermat 2002; 13(2):80-2.
58. Castelain M, Birnbaum J, Castelain PY et al. Patch test reactions to mite antigens: A GERDA multicentre study. Groupe d'Etudes et de Recherches en Dermato-Allergie. Contact Dermatitis 1993; 29(5):246-50.
59. Darsow U, Vieluf D, Ring J. Atopy patch test with different vehicles and allergen concentrations: An approach to standardization. J Allergy Clin Immunol 1995; 95(3):677-84.
60. Imayama S, Hashizume T, Miyahara H et al. Combination of patch test and IgE for dust mite antigens differentiates 130 patients with atopic dermatitis into four groups. J Am Acad Dermatol 1992; 27(4):531-8.
61. Vicenzi C, Revisi P, Guerra L. Patch testing with whol dust mite bodies in atopic dermatitis. Am J Cont Dermatol 1994; 5:213-5.
62. Beltrani VS. The role of dust mite in atopic dermatitis: A preliminary report. Immunol Allergy Clinics North Am 1997; 17:431-41.
63. Ring J, Abeck D. The atopy patch test as a method of studying aeroalergens as triggering factor sof atopic eczema. Dermatologic Therapy 1996; 1:51-60.
64. Pigatto PD, Bigardi AS, Valsecchi RH et al. Mite patch testing in atopic eczema: A search for correct concentration. Australas J Dermatol 1997; 38(4):231-2.
65. Mowad CM, Anderson CK. Commercial availability of a house dust mite patch test. Am J Contact Dermat 2001; 12(2):115-8.
66. Davis MD, Richardson DM, Ahmed DD. Rate of patch test reactions to a dermatophagoides mix currently on the market: A mite too sensitive? Am J Contact Dermat 2002; 13(2):71-3.
67. Jamora MJ, Verallo-Rowell VM, Samson-Veneracion MT. Patch testing with 20% Dermatophagoides pteronyssinus/farinae (Chemotechnique) antigen. Am J Contact Dermat 2001; 12(2):67-71.

INDEX

D

E

F

G

H

I

K

L

M

N

O

P

Q

R

S

T

U

V

X